Sick Planet

Corporate Food and Medicine

STAN COX

Pluto Press

LONDON • ANN ARBOR, MI

First published 2008 by Pluto Press
345 Archway Road, London N6 5AA
and 839 Greene Street, Ann Arbor, MI 48106

www.plutobooks.com

Copyright © Stan Cox 2008

The right of Stan Cox to be identified as the author of this work has been
asserted by him in accordance with the Copyright, Designs and Patents Act
1988.

British Library Cataloguing in Publication Data
A catalogue record for this book is available from the British Library

ISBN   978 0 7453 2741 9 (hardback)
ISBN   978 0 7453 2740 2 (paperback)

Library of Congress Cataloging in Publication Data applied for

10   9   8   7   6   5   4   3   2   1

Designed and produced for Pluto Press by
Chase Publishing Services Ltd, Fortescue, Sidmouth, EX10 9QG, England
Typeset from disk by Stanford DTP Services, Northampton
Printed and bound in the United States of America

CONTENTS

ACKNOWLEDGEMENTS

I am indebted to many, many people, both good friends and acquaintances, for ideas and information: Andrew Jameton, Kathleen Slattery-Moschkau, Allani Kishan Rao, Shailaja, Krishna, Manoj Agarwalla, Jenardhan, S. Jeevananda Reddy, Venkat Ram, Jayaram Mamidipudi, Marty Bender, Supi Seshan, Nabil Muhanna, Rosalyn Evans, Bruce Carraway, Maria Chavez, Manuel Chavez, Miranda Cady Hallett, Tim Crews, Y.V. Malla Reddy, and Kristan Markey. For inspiration, encouragement, and writing help I will always be grateful to Wes Jackson, Harris Rayl, Bill Martin, John Exdell, David Van Tassel, Chris Picone, Scott Bontz, Jonathan Teller-Elsburg, Tai Moses, Alexander Cockburn, Jeffrey St. Clair, Jan Frel, and finally, Roger van Zwanenberg, who gave me the chance to write a book. And my deepest thanks go to Greg Cox (my brother and chief medical advisor), Santosh Gulati (my mother-in-law, who kept me housed, fed, and encouraged during much of the writing), Tom and Brenda Cox (my parents, whose support is perfect and unwavering), Paul Cox (my son and research assistant), Sheila Cox (my daughter and first reviewer of anything I write), and Priti Cox (my wife, love of my life, and translator—even from English to English—who was with me every step of the way and without whom I could never have done this).

PREFACE

My aim in writing *Sick Planet* is neither to catalog the ecosphere's many grave symptoms nor supply a prognosis. You already know that the global outlook is grim and getting worse. Instead, I will show how ecological damage happens in two essential parts of our lives—health care and food—and argue that the changes needed to reverse that damage are much more radical than the dilute quarter-measures currently being proposed in Washington and other capitals.

Agriculture, food processing, and medicine generate smaller shares of total greenhouse carbon than do, say, personal transportation or indoor "climate control." But looking at the totality of life on Earth has led some of the key scientists in the mammoth Millennium Ecosystem Assessment project to the conclusion that agriculture plays a central role in "confronting the human dilemma."[1] And medicine, one of the fastest-growing sectors in industrialized economies, has become fully intertwined with food production. The stories in this book illustrate how all economic growth is ecologically destructive and why all of these sectors will have to be reined in together. Furthermore, it is crucial not to allow the biggest crisis looming ahead of us— rapid climate change—to blind us to other ecological problems that are already an everyday reality to impoverished people and threatened species on every continent. I can't make that case any more sharply than did Garret Keizer in the June 2007 issue of *Harper's* magazine. "Global warming, we are told, will have the most devastating effects on the world's disadvantaged," wrote Keizer. Therefore, according to too many experts, "we need not care so particularly about the world's disadvantaged; we need care only about global warming—as mediated, of course, by those who stand to make a bundle off it." Keizer concluded,

You do not repair the climate of an entire planet without staggering sacrifices unless the burden is shared with something like parity. To put that as succinctly as possible, the days of paradise for a few are drawing to a close. The game of finding someone else in some convenient misery to fight our wars, pull our rickshaws, and serve as the offset for our every filthy indulgence is just about up. It is either Earth for all of us or hell for most of us.[2]

It's too easy to see us all having a common interest in curbing climate change, whether we are tycoons or working people, whether we live in a powerful or a weak nation—to stress, in the words of former Vice-President and current climate-change ambassador Al Gore, that "we're all in this together."[3] True, but some of us are "in it" much deeper, and destined to sink much, much deeper, than others. And those divisions have everything to do with the causes of human-made climate change. The class struggle hasn't ended after all; it's going into sudden-death overtime. The global economy has proven itself capable of producing environmental misery and devastation at least as efficiently as it produces wealth. Those two faces of economic growth may be best illustrated by the ways in which the food and medical industries meet our fundamental biological needs. Each of this book's first nine chapters will illustrate contradictions in the way economic life-support systems operate in the world's second and third largest nations: India and the United States. More specifically, these chapters will show that:

- The fastest-growing major industry in the US, one meant to improve people's health, is instead undermining health—both directly and by degrading natural systems on which human well-being depends.

- Drug companies are achieving growth not only by making remedies for people who are sick, but also by creating whole new populations of sick consumers.

- The very symptoms meant to be treated by drugs manufactured in India for export are appearing among people exposed to pollution from bulk-drug factories there; even their ability to grow food is being lost.

- Nutritional products meant to cure health problems caused by overproduction and overconsumption end up stimulating greater consumption.

- As America mobilizes to protect industrial agriculture against terrorists, agriculture itself is doing the very kind of damage that terrorists are said to be planning.

- A "cleaner" fossil fuel increasingly being relied upon to curb global warming will be consumed at a rate that may threaten soil fertility and food production in countries already endangered by global warming.

- Rapid industrialization, being relied upon to pull a billion South Asians out of deep poverty, could end up weakening monsoon rains, making people even poorer and hungrier.

- Retailers are managing to sell more pleasing, healthful food, but almost exclusively to the well-to-do, and only by employing people who can't afford those very luxuries.

- Chemical compounds manufactured to help people cook in a more healthful way have been found in the bloodstreams of humans and other animals all over the planet—possibly causing cancer and other diseases.

There follows a tenth chapter that examines questions raised throughout the first nine, with help from three thinkers going back a century and a half: Nicholas Georgescu-Roegen, who demonstrated that all economic activity, whatever its purpose and however well it is done, inevitably accelerates the depletion of resources, production of waste, and sickening of ecosystems; Karl Marx, whose work showed how the essential mainspring of capitalism is the pursuit of insupportable growth; and William Stanley Jevons, who demolished the idea that resource efficiency alone can reconcile limitless growth with ecological sustainability.

The pages that follow will be populated with companies and individuals that are pushing the planet toward ecological ruin, but only as part of their routine, almost always legal, operations. And I won't be picking on especially bad examples. Sure, we'll encounter some scandal-ridden health-care corporations and meat-packing firms that have no qualms about ravaging their workers' health. But I'll go the hardest on those with seemingly good intentions, because they illustrate how idealism cannot tame capitalism's nasty side. I'll try to show that the planet's current predicament is not necessarily the work of evil, scheming tycoons bent on personal enrichment. Rather, it's the natural product of a system that rewards the industrious capitalist who pours a life's energy into building a vigorous, growing business in a competitive world. Just as we can't blame the current global predicament on "bad" corporate executives, we can't expect the "good" ones to come to the rescue. When corporate owners and managers claim they can't operate in greener ways without sacrificing essential profits, they aren't just being stubborn and greedy; they are acknowledging material reality.

The immediate causes of the destruction and misery that I'll describe may be industries or corporations or investors, but lying behind all of those is capitalist economics. If the human species, against all odds, finds an alternative to capitalism, it won't necessarily save the Earth. But if we find no alternative to capitalism, the Earth cannot be saved.

Efforts by "green" capitalists to pursue a so-called "triple bottom line"[4] by accounting for the well-being of people and nature along with profits, are as doomed as any effort to build a perpetual-motion machine. When those three goals come into conflict, as they inevitably will, it's the bottom-bottom line—profit—that must take priority. I also take no comfort in predictions that capitalism will erode its own foundations, eventually crumbling along with the breakdown of ecosystems and depletion of resources, ushering in a new, green era. John Bellamy Foster has argued convincingly that the astonishing ability of the capitalist system to adapt to almost any development, including environmental catastrophe, makes it all the more dangerous.[5]

The gold rush, or should I say carbon rush, stimulated by global warming is a stark example. As Foster points out, capitalism's internal contradictions—which impoverish workers who also happen to be consumers—create the kind of negative feedback that produces economic crises. But ecological destruction creates no similar feedback; a capitalist system can remain hale and hearty right up to the moment of utter calamity—as fully functional as a person who has just fallen the first 450 feet from the top of a 500-foot building.

Just as futile is the hope that increasing awareness of global climate change will wake people up to the danger of unlimited economic growth and, therefore, of capitalism. If the horrors that capitalism has inflicted over the past couple of centuries on billions of our fellow human beings have not shaken us out of our comfortable political positions, the threat of certain but only partially understood climatic disruptions won't do it either. That is why I will attempt in this book to demonstrate with specific examples the outrageous demands that capitalism places on humans and the planet. Then I'll ask you to widen your field of view from these stories of food and medicine to the global economy as a whole and imagine what we'll face if we continue to allow the growth requirements of capitalist economies to dominate over biological necessity.

I will warn you now that this book does not include a hopeful final chapter plotting a sure, safe route out of this mess. Before any new, ecologically sound society can be conceived, much less constructed, there has to be much wider agreement that the current economic system, with the engine of growth at its heart, cannot be part of that new society. My more modest goal, then, is to stir you to question both the desirability and the inevitability of capitalism on a sick, shrinking planet.

1

HEALTH CARE'S MALIGNANT GROWTH

In March 2006, a Lafayette, Louisiana cardiologist was indicted by federal prosecutors on 94 counts of fraud. Accused of performing unnecessary angioplasties, stent replacements, and other heart procedures, he was also hit with a civil suit by more than 300 patients who claimed they had been mined for profit.[1] A year later, as the cardiologist still awaited trial, a research report in the *New England Journal of Medicine* showed that many, perhaps most, of the million or so legitimate angioplasty and stent procedures done every year in the United States probably give little or no lasting benefit.[2]

On 23 January 2007, the Parker Hughes Cancer Center in Roseville, Minnesota filed for Chapter 11 bankruptcy. Its founder's state medical license had been revoked in 2006, three years after an investigative series by the *Minneapolis Star Tribune* revealed that the Center had been subjecting cancer patients to excessive testing and treatment. The paper accused Parker Hughes doctors of milking Medicare and private insurers by claiming that they could cure incurable malignancies and by having patients return as often as four times a week for unnecessary examinations and ineffective, often grueling, treatments. One surgeon who reviewed a typical case wrote to the state Medical Board that a patient's treatment had been extended "for profit, long after there was even the remotest chance for her obtaining any benefit from it."[3]

Also in January 2007, the state of Illinois joined a lawsuit against twenty magnetic resonance imaging (MRI) operators in the Chicago area for allegedly paying kickbacks to doctors who helped keep their machines supplied with patients.

Around the same time, federal prosecutors filed suit against a Florida radiologist for offering deals that could potentially net doctors or chiropractors as much as $30,000 per month in insurance payments so long as they referred enough patients for imaging procedures.[4]

Medical scams lure countless trusting patients into unnecessary treatment each year, but renegade doctors and clinics didn't create America's epidemic of overtesting and overtreatment. More imaging machines are kept humming every year, more blood vials are kept filled, and more hospital beds are kept occupied by good old respectable market forces.

For example, perfectly legal businesses like HealthFair USA and Life Line Screening offer ultrasound screening of arteries in the neck, abdomen, and legs, with fees running in the hundreds of dollars, depending on the number of arteries checked. The services can be arranged through websites, mail solicitations, or toll-free numbers advertised on the radio, or they can be done on the spot at "screening fairs" conducted via mobile labs. Faced with the popularity of ultrasound testing, federal and state public health authorities have advised people without known risk factors to avoid subjecting themselves to such mass screening programs.[5] False-positive results can turn healthy people into patients, subjecting them to further unneeded diagnostic procedures and possible treatment. Critics blame a host of other, more well-established tests, led by high-resolution mammography and the prostate-specific antigen test, for drawing throngs of healthy people into further, unneeded medical intervention.

"You can't live without it!" is one of advertising's oldest and most well-worn slogans, even though it's almost never literally true. But the medical profession intends to be taken literally when telling potential customers, "If you don't buy what we're selling, you run the risk of death or grave illness—and we can show you the evidence!" Nowhere, except maybe in the military-industrial complex, do business interests and life-and-death decisions intermingle as freely as they do in today's medical industry. And if there's any product that, to use another slogan of commerce, "sells itself," it's health care. That has helped it become America's premier growth industry, rising from 5 percent of the US economy in 1960 to 16 percent today.

"IF THEY BUILD IT, WE'LL FILL IT"

When the people and companies who do medical testing and diagnosis are connected to—or even the same as—those who do the treatment and surgery, the profitable opportunities multiply. *Time* magazine put it this way in a 2006 article on "The hospital wars": "Since physicians get paid through fee-for-service rather than, say, for curing their patients, their primary incentive is to do more stuff."[6]

And when doctors have a stake in diagnostic facilities, they have both the motive and the means to "do more stuff." For example, increasing numbers of doctors have been investing in their own MRI scanners, at prices ranging from $600,000 to $2 million apiece. Doctors, hospitals, and diagnostic centers did half a billion scans in 2006 alone. Some authorities are looking at the 40 percent increase in use of MRIs and other imaging procedures in the US since 2000 and wondering how many of those tests have been ordered solely for profit.[7]

Back in the 1960s, a California study showed that when physicians owned X-ray facilities, their patients ended up being X-rayed twice as often as patients whose physicians referred them to outside labs.[8] Three decades and several technological leaps later, the federal Government Accountability Office (GAO) studied records of almost 20 million office visits and found that doctors who had their own imaging equipment used the techniques two to five times as often as doctors who didn't.[9] Syracuse, New York has become what a 2004 *New York Times* article called the "unexpected epicenter for a high-tech medical arms race," with physicians throughout the city installing MRI machines in their offices; not unexpectedly, the number of scans shot up 23 percent in a single year.[10]

If you divide the price of a typical MRI machine by the typical fee a doctor collects for a scan, the result suggests that the doctor's first 2,000 to 3,000 scans will pay back the purchase price. Once that's done, additional scans generate what one health economist called "almost pure profit."[11] The proliferation of diagnostic equipment can have ripple effects that drive its use even higher. A recent review of the overuse of radiological testing concluded that "even in the absence of financial incentives, the mere availability of imaging technology in a nearby convenient location will lead to increased utilization."[12] As can be seen from Table 1.1, excessive testing has become a fact of medical life in America.

Table 1.1 Estimated percentages of diagnostic tests judged to be unwarranted.

Type of test, with reference	Percentage of the time that the test wasn't really needed
Imaging[13]	43
Urinalysis[14]	37
Biopsy or other invasive procedure[13]	32
Electrocardiogram[14]	9
X-ray[14]	7

That much superfluous testing is certain to result in a lot of unnecessary activity "downstream": more visits to doctors, more prescriptions written, more hospital admissions, more surgeries. One study found that if, say, 100 patients are each subjected to ten random diagnostic tests, around 40 of them will be "found" to have a problem that isn't really there.[15] A nationwide study of medical records by researchers at Georgetown University found large numbers of unwarranted urine analyses ordered for patients between 1997 and 2002.[16] Those tests would have led to as many as 28,000 unnecessary kidney biopsies, resulting in almost 1,500 cases of medical complications and thereby creating a new wave of patients with new problems.[17]

Asked by CBS News to respond to the Georgetown study, Robert Schwartz of the Miami University medical school cogently summed up the potential results of excessive testing:

It happens all the time ... The patient has no symptoms and doesn't smoke, but he gets a routine chest X-ray. If there is a small shadow, doctors are obligated to look further. That X-ray becomes a CT scan. That may show a small little nodule. The next thing you know, the patient ends up with a cardiothoracic surgeon who wants a needle biopsy, or even an open biopsy ... In a lot of these cases, he comes up with nothing, a benign nodule or something.[18]

Aggressive testing and treatment by a few for-profit medical facilities in a community can propel others in the same direction. The CEO of the four-year-old Green Clinic Surgery Hospital in Ruston, Louisiana told *Time* that competition from his clinic prompted a nearby regional hospital to respond with a "spending binge" that quickly ran the facility deep into debt, and into the arms of a new, for-profit owner. Meanwhile, the sprouting of physician-owned diagnostic and surgical centers across Wichita, Kansas has driven two big non-profit hospitals to invest in tens of millions of dollars' worth of new buildings and diagnostic equipment. Of that and other projects going on across the country, *Time* concluded, "If they build it, we'll fill it."[19]

Business columnist Steven Pearlstein of the *Washington Post* has formulated what he calls "Pearlstein's First Law of Health Economics," which states that "if you pay doctors on the basis of how many procedures they do, and you leave it to doctors and their insured patients to decide how much health care they get, consumption of health services will rise to whatever level is necessary for doctors to earn as much as the lawyers who sue them."[20]

Demand for more and more medical services is pumped up not only by excessive testing but also by a general "medicalization" of the human condition, as we'll see in the next chapter. Later chapters will examine ways in which other parts of the economy, led by agriculture and the huge industries that it feeds, is helping to create a nation of unhealthy people living in unhealthful environments—in other words, a permanent customer base for Big Medicine. And the health-care system has an unfortunate tendency to recycle its own customers: The

National Academy of Sciences estimated in 2000 that so many people are sickened or injured by medical error that each year between 44,000 and 98,000 patients actually die as a result.[21] Meanwhile, an estimated 4 percent of all hospital admissions are related to preventable problems caused by medication.[22]

If you're looking for scapegoats on whom to pin blame for the excesses of American medicine, there's no shortage of targets: malpractice lawyers accused of pushing doctors into "defensive medicine"; a private insurance system that pays for needless treatments; Medicare, blamed for inviting seniors to extra helpings of health care they wouldn't otherwise have or even need; patients who, it's said, will accept any treatment as long as someone else is paying; or those few "bad apples" running scams.

There is some truth to each accusation: Public and private insurance, the legal system, and our medicalized culture all really do work alongside overactive diagnostic labs to help feed medicine's out-of-control growth. But those aren't root causes; rather, they're a few prominent mechanisms among many by which the health-care system manages to snatch one dollar out of every six that flow through the economy. They all work together in the same direction—toward malignant growth—and they meet very little resistance along the way.

AN UNHEALTHY INDUSTRY

Debates about America's health-care crisis tend to focus on outrageous costs and the chronic lack of access for low-income and uninsured people to good-quality care. But flip that terrible problem over and you'll see its other face: a superabundance of medical services beckoning those patients who can afford them. As one observer put it, the system is suffering simultaneously from underuse and overuse.[23]

Access to adequate health care as a right of all the planet's people is recognized in the United Nations' 1948 Universal Declaration of Human Rights and subsequent international agreements. The United States stands alone among the world's wealthy nations in having no national policy guaranteeing affordable medical care for all. If someday we do manage to ensure that all Americans get all necessary health services, it will finally put an end to our national shame. But while delivering universal medical care would make the system more humane, it wouldn't make it sustainable. If that new system is also required to support the kind of huge corporate structure that the current system supports, and if the quantity of care to which people are entitled is not limited by something other than their income, the industry's environmental impact will grow worse rather than better. Like any overheated industry, American heath care is anything but environmentally benign. It's frittering away resources and throwing off wastes like, as the saying goes, there's no tomorrow. An industry dedicated to health ought not be feeding the endless economic growth that threatens the biological systems on which human health depends. But it is.

Big Medicine is a voracious eater of resources, as anyone who's been inside a well-furnished doctor's office or hospital in recent years can attest. In 2006, the US medical industry had $22 billion worth of buildings under construction or renovation— the biggest boom in half a century, predicted to last through the following decade. And despite a few environmentally friendly construction projects in recent years, the current hospital-building frenzy is coming down with the heavy impact that any construction boom tends to have. A report in the trade magazine *Health Facilities Management* summarized a nationwide survey of the "red-hot construction market that's reshaping the face of health care delivery." It extolled trends toward larger, more soundproof patient rooms, fewer shared rooms, nurses' computers in every room, wireless infrastructure alongside extra cabling and conduit, and of course, more and bigger electric power plants. But read through the report's 2,700-plus words, and you'll find not a single mention of energy conservation or other environmental issues.[24]

A hospital bed in America generates 8 to 45 pounds of waste every day, seven days a week.[25] That includes office paper, food, IV bags, gauze, syringes, human body parts, drugs, toxic agents used in chemotherapy, heavy metals, radioactive wastes, and much more. A typical pound of hospital waste contains three times as much plastic as does a pound of household trash.[26] Much of that plastic is polyvinyl chloride (PVC), which can leak toxic chemicals; without being aware of it, patients can be mainlining those toxins via intravenous drips. PVC can also emit highly carcinogenic dioxins when incinerated. Research is showing that many drugs, including chemotherapy agents, psychiatric drugs, anti-inflammatories and even caffeine, can pass, still in an active form, through bodies and into sewers and waterways. The sewer lines under hospitals and clinics are teeming with such compounds. Not all the drugs have passed through a kidney; unwanted or expired medications are often just dumped or flushed as well. Then there are "upstream" ecological costs; for example, the long, toxic history of a pair of latex or vinyl gloves that may be used for only a few seconds and discarded. Even way back in 1994, US hospitals were using 12 billion such gloves a year.[27]

The federal Centers for Disease Control and Prevention (CDC) estimates that 2 million people per year contract infections in America's hospitals, and that about 90,000 die from those infections.[28] As hospitals work to fend off the bacterial onslaught with disposable supplies and chemical disinfectants, autoclaving, and incineration, they chew up even more resources and spit out more wastes. By one estimate, "Treatment of the side effects of treatment accounts for about a third of all medical care."[29] And these days, detection and prevention of disease can be at least as resource-intensive as treatment. CT scanners, PET scanners, MRIs, and good old-fashioned X-ray machines are being joined every year by an alphabetful of new, improved contraptions. Diagnostic devices require huge computational power, heavily braced walls, vibration-resistant floors and/or lead shielding, and a big power grid on which to suck.

Andrew Jameton is a philosopher, a section head at the University of Nebraska Medical Center (UNMC) in Omaha, and one of the few academics willing to tackle the question of how to shrink medicine's big ecological footprint by shrinking the medical industry itself. In his office he has a simple diagram illustrating the cycle he sees driving the industry's rapid growth: "Big Medicine → Big Economy → Death of Nature → Poor Public Health → Big Medicine." As he took me on a backstage tour of UNMC, Jameton likened the typical American hospital's environmental wallop to that of a combination 24-hour hotel, truck stop, restaurant chain, office building, university science department, shipping company, and big-box retailer. He argued that there's an ethical imperative to rein in a system whose headlong growth seems to be producing more profit and less health. "But if you try to talk about ecological limits in the medical professions," he told me, "it's not a welcome conversation."[30]

In Jameton's view, medicine has outgrown sustainability because it's so lucrative for those who own it.

> Each year, we spend $5,500 to $6,000 per person in this country on health care. Who in the world can afford that? Everyone has to learn to live on less—and the rich will have to give up more than the poor. Does our own wealth relative to most of humanity entitle us to any treatment we demand, whatever the cost to the planet?

Jameton meant the question rhetorically, but the American health-care industry has an answer ready: Yes, if you're able to pay. The Kaiser Family Foundation has reported that in 2003, an astonishing 92 percent of all US health spending (both out-of-pocket and insurance-paid) was done by families in the top-50 percent income bracket.[31] That half of the population has fueled some phenomenal growth. Every year from 1999 through 2006, spending on health care grew at a faster rate than did the economy as a whole (Table 1.2). The 2001 recession that dragged down many other industries actually seemed to invigorate health care.

The federal Department of Health and Human Services (HHS) reports that per-person health-care spending has gone from $143 a year in 1960 to almost $5,700 today.[32] But complain as we might about medicine's soaring prices, the fortunes being made aren't all just the fruits of price-gouging. Removing the effect of rising prices from HHS's figures, we can calculate that the actual *quantity* of medical goods and services consumed per person has tripled since 1960.

Table 1.2   Annual growth rates (in percent) of US health-care
spending and the US economy as a whole, 1999–2006.[33] Whole-
economy growth is represented by growth in gross domestic product.
Rates are not adjusted for inflation.

Year	Health care	Whole economy
1999	7.1	4.8
2000	7.8	4.8
2001	10.2	2.1
2002	10.1	2.3
2003	7.8	3.7
2004	7.5	5.8
2005	7.4	5.4
2006	7.7	5.9

GROWING PAINS

Try as we might, our species will never manage to "destroy
the planet."[34] But we're doing a very good job of making it
sick. The most recent "ecological footprint" analysis by the
Oakland, California-based think-tank Redefining Progress
estimates that "on a global level, humanity is exceeding its
ecological limits by 39 percent. This suggests that at present
rates of consumption, we would need 1.39 Earths to ensure
that future generations are at least as well off as we are now."[35]
Meanwhile, the much-discussed 2005 Millennium Ecosystem
Assessment report, the work of more than 1,400 experts
worldwide over four years, concluded that 15 of the 24 key
"services" that ecosystems provide to humanity are in decline
because of human activity.[36]

Keeping the planet livable will require not just slower growth
of human economic activity, but an actual contraction. How
much will we have to cut back? Let's consider a few rough
estimates of the reductions that will be necessary in the US and
other Western nations. The Millennium Ecosystem Assessment
estimated that just to keep planet-wide temperatures from rising
more than two degrees Celsius (that being a threshold beyond

which research suggests that runaway heating will occur) requires that humanity's carbon emissions begin to decline by around 2015 and hold at 800 to 1,800 pounds per year per person worldwide by 2050.[37] The United States currently pumps out about 12,000 pounds of carbon per person, so to use only our fair share, Americans will have to slash their nonrenewable fuel use by 85 to 93 percent.

Considering not only carbon emissions but also the planet's overall health leads to similar numbers. The report on ecological footprints by Redefining Progress shows that per-capita consumption and waste production in Western Europe, if practiced worldwide, would reach almost four times the level that the planet could support. In the United States, the per-capita footprint is seven times the globally supportable level. To be a good world citizen and live on a fair share of world biocapacity, the average American would have to cut back by 86 percent, the average European by 75 percent.

If we accept that the carbon emissions and ecological footprints of Americans will have to shrink by 85 to 95 percent, how much less energy will we be able to use? The cornucopian concept being promoted by some of the high-profile environmental organizations in Washington—that we can cut emissions by 80 percent and still supply our current, voracious level of energy consumption solely through renewable energy—is a very bad joke.[38] The carbon threat is stampeding too many environmentalists into endorsing nuclear power and soil-destroying biofuel production as necessary evils. And no matter how much truly renewable energy capacity is developed, the economically and politically powerful interests that control oil, gas, and coal will see to it that those resources continue to be burned at the requisite pace. They will attack any effort, no matter how popular, to lock rich energy sources permanently in the ground. Their economic and political power will be at least as important as technological obstacles in holding down renewables' share of the total energy supply. But let's be perhaps overly optimistic and assume that half of our energy and materials consumption by 2050 will come from totally renewable sources. Then if we are going to live on 5 percent of our current emissions, we'll be able to use 10 percent as much energy as we do now. If we can tolerate 15 percent of current emissions, we can use 30 percent as much energy as now.

So consumption of energy and other resources will have to be reduced by 70 to 90 percent; split the difference and call it 80. And that relies on another pie-in-the-sky assumption: that future economic growth can be accomplished without adding to carbon emissions or other ecological damage. Furthermore, by 2050, we are likely to have added another 100 million to the current US population of 300 million. Taking everything into account, we'll be lucky if we can make do with only 80 percent less economic activity in the US and still have a healthy planet.

No current plans for ecological sustainability consider the necessity, let alone the possibility, of shifting economies into reverse gear. Even the most environmentally conscious prescriptions focus on "smarter" or slightly slower growth. But a deep contraction of human economic activity is never, ever discussed by the global economy's decision-makers, because they understand only too well their utter dependence on ever-expanding consumption.

In the history of capitalist economies, various sectors have taken turns shouldering the burden of stimulating consumption. In the US over the past decade, through good times and bad, it has fallen to the health-care industry to carry a big share of that burden. Because it has been one of the US economy's few consistent job-generators, and because its declared mission is to save and improve lives, Big Medicine's rush to grow has not been seriously addressed by the people and corporations that own it, the doctors and other health workers whose salaries and wages it pays, or the patients it treats.

Health care is currently serving better as an economic than a biological life-support apparatus. In mid-2006, the publication *Business Week* starkly highlighted the crucial role that the medical industry (including health insurance) has come to play in the US economy.[39] From 2001 to 2006, medicine was credited with creating 1.7 million new jobs. The net number of jobs created by the rest of the economy during those years added up to precisely ... zero. Were the economy to keep to its present course, 30 to 40 percent of all jobs created over the next quarter-century would be in health care. The magazine noted that without the health-care industry, "the nation's job market would be in a deep coma."

But Big Medicine is no mere make-work program. What was once a public service has become a high-horsepower wealth engine. While profits rose by just 5 percent from 2000 to 2004 for US corporations in general, they more than doubled for the largest health-insurance firms, whose stock value also doubled.[40] The rate of profit in the pharmaceutical industry averaged an exuberant 17 percent from 1999 to 2004, never falling below 14 percent.[41]

THOSE BAD APPLES

Wherever you find breakneck growth, you find at least a few companies moving ahead of the competition, into questionable territory. The rise to dominance of for-profit health-care corporations since the 1980s has been punctuated by some spectacular implosions of companies whose growth took them a bit too far ahead of the competition. As Pulitzer Prize-winning reporters Donald Bartlett and James Steele document in a key chapter of their 2004 book *Critical Condition: How Health Care in America Became Big Business & Bad Medicine*,[42] Wall Street's top investors steadfastly backed each of those renegade corporations, from their rise to their fall and sometimes beyond.

The Wall Street investment house Smith Barney and the Swiss bank UBS Warburg helped a 1984 Birmingham, Alabama startup called the HealthSouth Corporation grow into the nation's largest outpatient and rehabilitation service company in the late 1990s. But by 2002, in Bartlett and Steele's words, "the company had been cooking the books through a scheme that inflated earnings to meet Wall Street's projections ... It was a Ponzi scheme of sorts ... Health South's felonious accounting gave new meaning to the concept of playing with the numbers, according to SEC and shareholder lawsuits."[43]

But UBS Warburg and other financial backers continued to express their confidence in HealthSouth right to the bitter end. They claimed to have been deceived like everyone else, but after HealthSouth finally crashed and a federal court indicted its founder Richard Scrushy in November 2003 on 85 counts of fraud, money laundering, and conspiracy, investors wasted no time in finding new growth opportunities in Big Medicine. As Bartlett and Steele wrote, "Wall Street never looks back. It simply moves on to the next target in a never-ending search to feed its perpetual-motion machine of fees, IPOs, debt offerings, and restructurings."[44] In 2005, Scrushy was acquitted of the charges related to HealthSouth's operation. Columnist Michael Kinsley, with tongue in cheek, attributed the acquittal to divine intervention.[45] But the fact that Scrushy made, and was allowed to keep, almost $300 million from HealthSouth's allegedly fraudulent activities is simply further evidence that the US economy has a very high tolerance for reckless growth. In 2006, Scrushy was convicted and received a federal prison sentence for contributing $500,000 to a former Alabama governor in exchange for a seat on a state hospital licensing board. That same year, HealthSouth regained its listing on the New York Stock Exchange.

Just before Halloween, 2002, the Federal Bureau of Investigation raided Tenet Healthcare Corporation's Redding Medical Center in California, where, it was charged, two doctors had been running a "cardiac factory," performing large numbers of unnecessary artery-bypass and heart-valve surgeries. Lawsuits over what was called a "scheme to cause patients to undergo unnecessary invasive coronary procedures" cost Tenet and the doctors a total of $370 million in settlements over the next three years. Throughout the decade-long ascendance of Tenet to its eventual position as the nation's number-two for-profit hospital chain—through the disclosures of excessive treatment-for-profit, horror stories of patient infections caused by cost-cutting on sterilization, outrageous executive salaries and bonuses, executive antics at extravagant Las Vegas conventions, and warning signs of the ignominious collapse that would come in 2003—"Wall Street continued to love everything about Tenet, always finding something to praise," according to Bartlett and Steele.[46]

The antics of HealthSouth, Tenet, and other corporations were not aberrations of the US health-care system; rather, those now burnt-out companies served as scouts on the frontier, leading the industry's expansion. If investors were willing to lend such enthusiastic support to such ruthlessly profit-oriented corporations, they can also be counted upon to put dollars ahead of health when backing those more reputable industry players who manage to grow unhindered by indictments or bankruptcy.

Bartlett and Steele's book was primarily concerned with the chaotic behavior and high prices that privatization has brought to American health care in recent decades. But the economic forces they documented have also helped to bring fantastic growth in the system's physical bulk and environmental impact.

GREEN HEALTH CARE?

A growing number of professionals have recognized that Big Medicine's assaults on nature are helping undermine the very human health that the industry is counted upon to protect. That has led to formation of several large national organizations dedicated to "greener" medicine, and to some partial successes on specific issues:

- A study by the American Hospital Association and Hospitals for a Healthy Environment found that 80 percent of hospitals surveyed had stopped using fever thermometers containing the highly toxic element mercury.[47]

- The list of cities and organizations formally aiming to eliminate PVC, dioxins, and/or waste incineration in medical facilities is lengthening.[48]

- The company PharmEcology Associates is working with some success to reduce drug pollution flowing from medical facilities into sewer systems.

- Groups such as Sustainable Hospitals have developed highly detailed guides to "environmentally preferable purchasing." The Nightingale Institute mobilizes nurses and clinicians to push for more environmentally sound products and procedures in their own workplace.

- A campaign by Hospitals for a Healthy Environment and the Green Guide for Health Care is promoting eco-friendly hospital guidelines that include 96 design and construction features and 72 operation principles.[49]

Ted Schettler is science director of the Science and Environment Health Network. Although, he told me, "There's plenty of work yet to be done," he has been pleased to see a growing list of hospitals strive to reduce or eliminate mercury, PVC, waste incineration, and drug-dumping. And he's encouraged by a trend in some areas toward green medical buildings. But, welcome as they are, green innovations being adopted by hospitals remain peripheral to the main problem of ballooning growth. When I asked Schettler about anti-Big Medicine philosopher Andrew Jameton's argument that any environmental gains achieved by using better materials and methods can easily be eaten up quickly by an industry that at its current growth rate will double in size in less than two decades, his upbeat feeling seemed to fade. Schettler knows Jameton and agrees with his analysis, but, he said, "That's a tough one. People are not going to give up access to expensive medical care."[50]

In 2004, Jameton and Jessica Pierce, lecturer in philosophy at the University of Colorado, Boulder, co-authored a book entitled *The Ethics of Environmentally Responsible Health Care*.[51] In it, Pierce and Jameton described a hypothetical "Green Health Center" that would, they argue, achieve better medical results more cheaply and with a lower ecological impact. Despite evidence that they would dramatically reduce costs and turn out healthier patients, there are no actual Green Health Centers in operation or under construction. A review of Pierce and Jameton's book in the British medical journal *The Lancet* praised the authors for taking on the challenge of defining true sustainability in a medical facility but wound up dismissing the Green Health Center idea as economically impractical. *The Lancet*'s alternative suggestion was less than inspiring: "At this juncture, we need simple, tentative, precautionary approaches that cut through the uncertainties revealed by science. We need to buy time to find smarter ways of living while not crippling our economies in the process."[52] In other words, tidy up around the edges, but never interfere with economic growth.

Pierce, the book's lead author, told me, "We wrote it as a utopian vision, and we hope health care will evolve toward that vision. But we really are presenting a pretty serious challenge to the economic structure." In her view, the ecological damage caused by medicine has grown out of a badly distorted growth in its priorities. "The crux of our argument is that allocation of our spending is misplaced. In the past, the greatest advances in overall health have come from clean water, clean air, public works, public health, preventive care."[53]

In a more recent article, Jameton outlined a comprehensive approach to cutting US health care's energy consumption by 80 percent. He recommended that the nation:

- First make a 50-percent cut by ranking all health services and eliminating those that are most "useless, harmful, unnecessary, wasteful, or costly." Jameton claims the support of much of the medical community in arguing that such a 50 percent cut would actually improve Americans' overall health.

- Extend access to people who currently cannot afford health care, through public health programs and greater income equality. That would constitute a necessary setback in energy conservation, canceling out about a third of the savings from the 50 percent cut in services.

- For dying or incurably ill patients, "focus less on devices to cure, or to pursue the illusion of cure, and more on high quality nursing care and comfort."

- Restore and improve public health services that prevent illness and injury.

- In medical research, "instead of focusing on finding new diseases to treat and new therapies for them, we undertake an intensive effort to reduce the energy costs of caring for and curing people."

- Reduce the "background costs" that feed the medical system, through improved conservation in electricity generation, transportation, manufacturing, construction, etc.[54]

A survey published in 2007 by a group of Australian and British researchers asked 253 health professionals in Indonesia, Thailand, India, Iran, South Africa, and Bulgaria to imagine that they were in charge of allocating funds for health care in their respective countries and to rank ten types of medical intervention in order of priority. The three items receiving highest scores were child immunization, anti-smoking education, and general practitioners' care for everyday illness. Lowest scores went to complex surgical procedures.[55]

In this country as well, simple health-preserving strategies would undoubtedly be more effective than what we are doing. Yet only about 2 to 3 percent of US health-care spending is for preventive care, because mundane prevention measures never produce the splendid profits that high-tech medicine delivers. In searching for guidance to improve its efficiency, Big Medicine has even turned, believe it or not, to automobile manufacturing. A commentary in the *Journal of the American Medical Association* entitled "Reducing waste in US health care systems" described in detail how Toyota's "lean production" methods can be applied to health care.[56] In their book, Bartlett and Steele describe how management consultants visited a Saturn automobile assembly plant in the 1990s "looking for tips on labor-saving techniques and processes on a manufacturing assembly line that might be adopted by hospitals."[57]

But instead of more sophisticated medical care, Jessica Pierce told me, "We need more 'human care,' before people ever get sick.

As it is, the system is undermining the very health it's supposed to be protecting. And a lot of those treatments and technologies have negligible health impacts." She drew her own automotive comparison: "People have a choice to buy a Hummer, too. That doesn't mean society should encourage them to do it."

THE MODEL WEALTH CREATOR

Today's economy pushes people not only to buy Hummers but also to accept every medical product and service that might have even the slightest chance of benefiting them. There's too much at stake to think of cutting back. US economists are deeply grateful to the medical industry for keeping the nation's gross domestic product (GDP) pumped up, but as even the awe-struck *Business Week* article on the subject acknowledged, "That sort of lopsided job creation is not the blueprint for a well-functioning economy."[58] That's a rare admission in mainstream circles, where anything that helps the GDP grow is welcomed with open arms, because it means that somewhere, someone's making money—even if someone has to get sick or die to make it happen. Friends of the Earth UK has put the situation in the bluntest terms, noting that "a cancer patient is a model GDP-wealth creator."[59]

Ecological economists and other heretics of the profession have long pointed out that the GDP is a lousy indicator of economic well-being, because it lumps together desirable and undesirable goods and services—putting everything in the "plus" column—while ignoring important and useful activities that don't cause money to change hands. Alternatives to GDP such as the Index of Sustainable Economic Welfare (ISEW) and Genuine Progress Indicator (GPI) attempt to account for desirable goods and services while subtracting undesirable "bads."[60] While the size of the US economy in the form of per-capita GDP has marched steadily upward over the past century, the actual well-being of the US population expressed as GPI has been drifting along at an almost constant value since about 1970.[61]

Individual medical goods and services are not easy to classify as good or bad. Clearly, saving a life is good and necessary, but one frantic evening in the emergency room is not an isolated event. Medicine is embedded in a vast web of mutually reinforcing economic activities that, as we will see in coming chapters, are simultaneously creating and solving problems, while generating new capital at both ends. Publicly supported hospitals are as entangled in that web as are the big, greedy health-care and insurance corporations. The prowess of agriculture in producing food surpluses; the ever-more-generous portions served up by the restaurant industry; the transportation sector's insistence that its survival depends on high speed limits; the home-entertainment industry's amazing ability to keep able-bodied boys and young men motionless, except for their thumbs, for hours at a stretch; the chemical manufacturers that have infiltrated scores of synthetic compounds into the human bloodstream; the advertisers of tobacco, alcohol, and other legal drugs, whether over-the-counter or prescription; the bosses who have added a full month of work and stress to the typical working couple's year in the past decade – each of these GDP-lifting branches of the economy both feeds and is fed by the health-care industry. Even if accountants could quantify these and all the other not-so-good goods and services, it seems impossible to figure out where to begin extracting them from the web.

However, we can tug at some of that web's fibers. Next we'll follow an especially long, thick one: the prescription drug trade.

2

FEELING OK? ARE YOU SURE?

When you're watching a hospital drama on TV and start to feel a bit unwell yourself, a psychologist might diagnose your reaction as a type of "empathetic identification." When you see a commercial for a prescription drug and start feeling symptoms— a twinge in the leg or maybe a moment of doubt about your emotional stability—that's called effective advertising.

Spending by pharmaceutical companies to promote their products in the US (through advertising, gifts to doctors, sample giveaways, and other means) doubled between 1999 and 2004. Spending on direct-to-consumer advertisements for prescription drugs more than doubled during that same period,[1] and the tide of TV commercials continues to rise. Attempts to shut down direct-to-consumer ads have so far failed, and the spots continue to do their job very well indeed. Each year, the drug industry churns out enough products to fill more than 15 prescriptions per American,[2] and that adds up to more than $200 billion in annual sales.[3]

It seems that everyone who's not a pharmaceutical executive has a gripe about rising drug prices, but price-gouging doesn't completely account for the companies' hefty profits. They depend just as much on growing markets. One way to expand the market for medicines is to expand the numbers of people who are, or believe themselves to be, ill. To that end, as all Americans are by now painfully aware, the companies and the "awareness" groups they fund have defined and redefined a host of medical conditions—like erectile dysfunction, female sexual dysfunction, restless legs, sleeplessness, bipolar disorder, attention deficit disorder, social anxiety disorder, and irritable bowel syndrome—to include larger and larger segments of the population in the United States and other Western nations. To accept industry's claims about the numbers of people suffering from the eight diseases listed above would be to accept that more than 90 percent of adult Americans have been struck by at least one of them.[4] Throw in a few more conditions like depression, elevated cholesterol, and bone-density loss, and widely advertised affliction figures make it appear that virtually every American has a disease in need of a treatment. Those are all real maladies for some people, but the name of the medical game for the past decade has been to publicize inflated numbers in order to swell the customer base for profitable drugs.

DISEASE MONGERING

Lynn Payer's 1992 book *Disease Mongers: How Doctors, Drug Companies, and Insurers Are Making You Feel Sick*[5] alerted the US public to the pharmaceutical industry's tricky tactics. Thirteen years later, Ray Moynihan and Alan Cassels' book *Selling Sickness: How The World's Biggest Pharmaceutical Companies Are Turning Us All Into Patients*[6] showed how the industry's promotional stratagems have been refined and amplified. Indeed, the companies could soon achieve the astonishing feat of medicating an entire population. Here are just a few of the highly profitable medical conditions that have been marketed:

Restless legs. The evolution of "restless legs syndrome," as outlined in a 2006 paper by Steven Woloshin and Lisa Schwartz,[7] is a case study in how a pharmaceutical company, with help from the media, can turn what is a serious problem for some people into a contrived medical condition for millions more. Woloshin and Schwartz analyzed media coverage in the interval between 2003, when drugmaker GlaxoSmithKline, Inc. first issued press releases about trials of its drug Requip for relief of restless legs syndrome, and 2005, when the US Food and Drug Administration (FDA) approved Requip for that use. Of 187 major newspaper articles published during those two years, 64 percent relayed without comment the industry's claims that millions of Americans—as many as "1 in 10 adults"—suffer restless leg. Forty-five percent of the articles stressed that many people may be unaware they're sick, even though, according to 73 percent of the articles, the syndrome can have extreme physical, social and emotional consequences. Reports of the relief provided by drug treatment used "miracle language" 34 percent of the time, while 93 percent of articles failed to quantify Requip's side effects. Defending his company in 2006 against charges of "selling" restless leg, David Stout of GlaxoSmithKline told the press, "You need to talk to the patients. Things like restless leg syndrome can ruin people's lives. It is easy to trivialize things when you don't have them. If people did not want the treatments, they would not seek them."[8]

Restless leg syndrome in its most serious form is indeed no joke. My own father, for one, was tormented for years by its symptoms, until, without ever having seen an advertisement, he sought treatment. But, says Dr. David Henry, physician and professor at the University of Newcastle in Australia, "When you extend a drug beyond the most severely afflicted group on which claims of its effectiveness are based, you see a falling ratio of good to harm. The benefits of the drug diminish, while the side effects tend to stay the same. Companies know quite consciously that they're going into areas where they're doing net harm."[9]

Woloshin and Schwartz pointed to restless legs syndrome as a prime example of those "disease promotion stories" that the press loves to cover: "The stories are full of drama: a huge but unrecognized public health crisis, compelling personal anecdotes, uncaring or ignorant doctors, and miracle cures." The story of another disease, irritable bowel syndrome, has all of those dramatic elements—plus dead patients.

Irritable bowel syndrome. In *Selling Sickness*, Moynihan and Cassels describe public-relations offensives that were launched by Novartis Pharmaceuticals and GlaxoSmithKline to popularize a condition called irritable bowel syndrome,[10] symptoms of which are described as "abdominal pain or discomfort associated with changes in bowel habits in the absence of any apparent structural abnormality."[11] The companies stood to gain billions in sales if, as they claimed, as many as 20 percent of Americans had the syndrome. GlaxoSmithKline's drug Lotronex received FDA approval for treatment of irritable bowel in 2000, and Novartis' Zelnorm was approved in 2002. In statements to the FDA and the public, the companies tended to describe irritable bowel syndrome as it is experienced by the worst-afflicted patients—a tiny percentage of the total—while emphasizing claims that the syndrome hits vast numbers of Americans.[12]

The FDA wrote to Novartis in 2003, demanding that the company discontinue advertisements that it considered misleading. According to Moynihan and Cassels, the drug came under fire in late 2000 when three FDA scientists wrote to their superiors expressing alarm over a rising toll of deaths and hospitalizations of irritable-bowel patients during the nine months that Lotronex had been on the market.[13] Today, FDA requires that Lotronex be prescribed only by doctors who have enrolled in a GlaxoSmithKline "Prescribing Program."

Attention deficit disorder. A more widely discussed disease, attention deficit disorder (ADD, also called attention deficit hyperactivity disorder, ADHD), is responsible for an enormous prescription market among young people; for example, the National Institute of Mental Health estimates that there's an average of at least one afflicted child per typical-size classroom. But people spend many more years as adults than as children, and stiff competition among the major ADD drugmakers—among them Shire PLC, Novartis, and Eli Lilly and Company—guaranteed that sooner or later the larger pool of potential adult patients would be tapped.

The marketing of ADD can venture into bewildering territory. Lilly earned a 2006 "Bitter Pill Award" from the Prescription Access Litigation Project for its TV commercial plugging the ADD drug Strattera. In the ad, the legally mandatory information on approved uses, risks, and side effects is accompanied by wildly distracting sights and sounds of a video game. The FDA issued Lilly a mild rebuke over the ad, stating, "The overall effect of the distracting visuals and graphics is to undermine the consumer's ability to pay attention and comprehend the risk information...."[14] The Bitter Pill Award citation stressed the obvious irony of an attention-confounding ad targeted at a clientele who are believed to have difficulty paying attention. The Strattera campaign also won Lilly a 2005 Pharmaceutical Advertising and Marketing (PhAME) Award, this one in praise, not mockery.[15]

As with many afflictions, standard advertising and other direct promotion of ADD is supplemented by the work of a nonprofit "awareness" group. Children and Adults with Attention Deficit/Hyperactivity Disorder (CHADD), which gets about 20 percent of its funding from drug firms, energetically promotes the idea of ADD as a lifelong condition. The group's website provides ample and detailed advice on medication for ADD. One example:

> Although there is little research on utilizing short-acting and long-acting medications together, many individuals, especially teenagers and adults, find that they may need to supplement a longer-acting medication taken in the morning with a shorter-acting dose taken in mid to late afternoon. The "booster" dose may provide better coverage for doing homework or other late afternoon or evening activities and may also reduce problems of "rebound" when the earlier dose wears off.[16]

Sexual dysfunction. Seeing the sustained flood of advertising for impotence remedies in the American media, a visitor from the planet Zefitor could be forgiven for wondering how, with such dysfunctional human males, planet Earth ever came to be inhabited by 6.6 billion of the species. The erectile-dysfunction advertising blitz has its roots in the 1999–2001 Pfizer, Inc. campaign that transformed the father of all impotence drugs, Viagra, "from an effective product for erectile dysfunction due to medical problems, such as diabetes and spinal-cord damage, into a drug that 'normal' men can use."[17] Pfizer spent $303 million in direct-to-consumer advertising in those years for Viagra, often featuring younger-looking men and sports stars. That effort paid off handsomely, by extending the market well beyond men with well-defined medical conditions and attaining its greatest sales growth in the 18 to 45 age group.

Pfizer's salesmanship broke the age barrier for Viagra, but the company failed to extend the drug's market to that half of the human population that is immune to erectile dysfunction—that is, to women. According to Leonore Tiefer, coordinator of the Campaign for a New View of Women's Sexual Problems, the term "female sexual dysfunction" (FSD) traces back to 1997. In the years that followed, demand for a "pink Viagra" was boosted by sisters Jennifer and Laura Berman, who, Tiefer has written, "became the female face of FSD, opening a clinic at the University of California Los Angeles (UCLA) in 2001, and continuing to popularize FSD and off-label drug treatments on their television program, website, and books; in appearances on the television show *Oprah*; and in innumerable women's magazines."[18] Pfizer aggressively promoted FSD, which it labeled "female sexual arousal disorder." But its plans to provide a Viagra for women eventually fizzled, according to Tiefer, because of poor clinical-trial results.

The above examples hardly exhaust the range of strategies used in disease mongering. The ranks of sufferers of bipolar disorder, which, based on its original diagnostic criteria, afflicts a tiny 0.1 percent of the US population, have been swelled by a dubious re-definition that includes as many as one out of every 20 Americans.[19] Antidepressants are thought to be widely overprescribed; certainly, their use continues to grow.[20] One of the more memorable cases of depression-mongering was GlaxoSmithKline's marketing of a condition called "social anxiety disorder" (conveniently abbreviated "SAD"), which succeeded in making the company's drug Paxil the nation's leading antidepressant for a time in 2000.[21] And blood-cholesterol levels that trigger drug treatment were lowered, first in 2001—a change that tripled the numbers of people who qualified for treatment[22]—and again in 2004.[23] Critics charged that eight of nine authors of the 2004 revised guidelines had financial ties to the cholesterol-drug industry and that the American Heart Association, which endorsed the new guidelines, receives funding from the industry.[24]

A 1994 World Health Organization (WHO) report defined "normal" bone density as that of a healthy 30-year-old woman. Because virtually all women experience natural loss of bone density as they age, the WHO criterion led to the creation of a new medical condition called "osteopenia," which is said to afflict 30 percent of all American women over the age of 50. That's in addition to the smaller number of mainly older women who have the serious bone loss disease called osteoporosis. Women who have passed menopause are now under tremendous pressure to have bone scans done, greatly increasing the potential customer base for anti-osteoporosis drugs. Although hip fracture is the greatest danger posed by the more serious, established disease osteoporosis, mass bone scanning does not predict very well at all the risk that a patient will eventually break a hip; the best indicator, not surprisingly, is simple age.[25]

Finally, anyone who managed to stay awake through network TV commercials during 2006 and 2007 can identify the top disease-mongering target of that period: sleeplessness. The $345 million being spent annually by drug companies on ads for sleep aids was supplemented by a federal study (conducted partly at the companies' request) purporting to show that 50 million to 70 million Americans suffer from sleep problems and that US businesses lose as much as $100 billion a year because of sleepy workers.[26]

DIRECT-TO-PATIENT, DIRECT-TO-DOCTOR

The University of Newcastle's David Henry, who has published and organized symposia on the problem, is pessimistic about turning the tide of disease mongering. He told me:

It can't be stopped. It's a consequence of our political economy, the domination of marketing in all areas of life. That means we need to build counterforces. People are becoming more skeptical, and that needs to be

encouraged. We should exercise the same healthy skepticism when being
sold a drug as we do when being sold a secondhand car.

He says greater use of the attention-getting term "disease
mongering" will prove useful in changing the behavior of
medical professionals, the media and even pharmaceutical
public-relations departments. "We want it to be an idea that
pops up in their heads, so PR people will say, 'Hey, we don't
want to run this ad and be accused of *disease mongering*!'"[27]

But it will doubtless take more than vivid language to curb
the limitless profit hunger of the drug companies. A proposal
to end direct-to-consumer television ads has been introduced in
Congress,[28] and federal regulators are making more moves to
crack down. But for a corporation seeking new customers, there
are always alternatives. Spending on internet ads for prescription
drugs is predicted to reach $1.3 billion by 2008, exceeding what
was spent on *all* direct advertising as recently as 1998. That was
the estimate of eMarketer.com, which noted that the internet
is now the first source of health information for more than 30
million Americans. The report's author pointed out, "The result
is a shift in focus from direct-to-consumer to direct-to-patient,
from mass marketing to relationship marketing."[29]

That trend will continue to be paralleled by the relentless
pressure that pharmaceutical companies' sales representatives
exert directly on doctors, pushing them to keep dosing patients
with the "best" drugs. Despite the rapid rise of direct-to-consumer
advertising, 90 percent of the pharmaceutical companies' $21
billion annual marketing budget (more than ten times the entire
FDA budget) is directed at influencing doctors, largely through
sales reps.[30] It's widely believed that reps apply that pressure in
order to displace a competitor's brand of drug with one of their
own that has the same purpose. Companies also draw attention
to situations in which drugs can treat conditions that otherwise
would require surgery. But relationships between sales reps and
doctors, like disease-mongering commercials, also work by
inflating the total number of prescriptions being written. For a

patient, every prescription is another gateway into the labyrinth of Big Medicine.[31]

Kathleen Slattery-Moschkau was a pharmaceutical rep for ten years in the Madison, Wisconsin area, working for Bristol-Myers Squibb Company and Johnson and Johnson Pharmaceuticals, Inc. She went on to write and direct a feature film *Side Effects* and a documentary *Money Talks: Profits Before Patient Safety*, both released in 2005 and both exposing the not-so-pretty strategies and tactics drug companies use in dealing with physicians. In her view, reps are under pressure not necessarily to see that patients get well but rather to grab a bigger market share for the rep's company—or simply to create a bigger market. As Slattery-Moschkau puts it, "If you make the pie bigger, every company's slice is bigger, and everyone's happy."[32]

A survey of 29 published academic studies showed that typical interactions between sales reps and doctors—including gift-giving, providing free drug samples, paying for meals and travel, speaking before groups, and providing research funds—all have the desired effect. Targeted doctors prescribed more of the drugs in question or requested more often that they be added to pharmaceutical stocks at the hospitals or clinics where they practiced.[33]

I asked Slattery-Moschkau what life as a drug rep was like. She said:

> We were expected to see at least 8 to 10 physicians a day. Company statistics showed conclusively that the more of those face-to-face meetings that happened, the higher the sales would be. A rep's day-to-day job mostly involves schlepping around a lot of things like drug samples and food. All I knew about medicine was what the company told me. I was a political-science major.

But the company had exactly the information she needed:

> We knew how many prescriptions each doctor was writing for each drug; the company provided us the data. So we knew who were the biggest "writers" for, say, a particular anti-depressant. We were told, "Don't even pay attention to the ones in the lower tier of writers." And it wasn't just

prescription data; we maintained personality profiles on doctors: "Is this one more analytical? Do I need to take him some published studies? Is that one more susceptible to a dinner out?"

In 2007, Vermont became the third state to prohibit companies from gathering and distributing data on the prescribing habits of individual doctors.[34] But in the rest of the country, drug reps continue to rely heavily on detailed information about doctors' writing histories.

The "key way" of getting more prescriptions written, said Slattery-Moschkau, was to encourage "opinion leaders"—that is, well-respected doctors—to "use, believe in, and talk about our products. It's ten times as easy to sell a drug if a top dog is writing prescriptions for it." All reps would arrange "educational speaking engagements" in their respective sales territories for high-profile specialists: "We'd pay their travel expenses and speaking fees. We'd have a prominent medical expert pitching our message, twisting the drug information just enough, focusing on the positive and leaving out most of the negatives."

Doctors in America are generally very well-off and don't need someone to pay their restaurant tabs or buy football tickets for them. But research has shown that all gifts, regardless of size, have an influence on prescribing behavior.[35] The companies clearly know that. When hosting an influential speaker, says Slattery-Moschkau,

> We'd give the event an "educational"-sounding title and hold it at an expensive restaurant, or maybe even have the speaker do the presentation on the way to a Packers game. Two weeks later, we'd get the numbers on whether the doctors who'd attended were writing more, and we'd know if it worked. If it did, if the opinion leader's talk had driven up market share, the word would spread like wildfire among the reps, by voice mail, email, word-of-mouth: "Hey, this guy moved us 10 percent with just one talk!" Or if there were no results, the message would go out: "Never use this guy!"

In 2002, the Pharmaceutical Research and Manufacturers of America (PhRMA) adopted a code (endorsed by the American

Medical Association, AMA) that was meant to restrict pharmaceutical reps' freedom to woo doctors.[36] A review of that code
in the journal *Business Ethics Review* pointed out that similar
attempts in the early 1990s to encourage voluntary restraint had
been unsuccessful, and the review's author held out little hope
that the new code alone could bring real change:

> But how do the new AMA Guidelines and PhRMA Code create an ethical
> professional and business environment that will be any more effective
> than the one that existed a decade earlier? Like their predecessors,
> they both remain voluntary standards of behavior; and while specific in
> nature, neither the AMA nor PhRMA require that their members abide
> by these ethical statements or risk sanctions, fines, or expulsion from
> their respective associations. Thus, enforcement of these professional
> and industry conduct codes is placed wholly on the individual or
> company management.[37]

I asked Kathleen Slattery-Moschkau if she thought the
guidelines had had much effect. "In short, no," she said.

> Industry people with whom I've stayed in contact, or who have seen my
> films say that maybe they can't fly doctors to exotic islands for a holiday
> anymore, but that they can continue most practices as long as they're
> in the name of education. When companies say, "Don't worry—we can
> regulate ourselves" and publish self-imposed guidelines, it's just public-
> relations spin.

Yet another attempt to reduce reps' influence came in the
form of a 2006 policy proposal written for the *Journal of the
American Medical Association* by authors from seven public
institutions. They recommended a strict ban on most rep–doctor
interactions, at least in academic hospitals and medical centers.
In deadpan language, the authors pointed out that pharmaceutical companies' "ultimate fiduciary responsibility is to their shareholders who expect reasonable returns on their investments."[38]
One mainstay of drug-company strategy, the distribution of drug
samples to doctors, was among those practices those authors
would like to see prohibited.

That would surely meet fierce opposition. Many doctors or medical students who've attended screenings of her films tell Slattery-Moschkau she's right about the pernicious influence of drug reps. "But," they tell her, "I'll still have to visit with the reps because of the samples. Some of my patients really need them." But that's hogwash, she told me: "Samples are nothing but a marketing tool. And they're so effective! It's nothing to give away a week's or a month's worth of pills that you know a patient's going to be on for a lifetime."

Samples become an even more effective weapon when part of a "study":

> Occasionally with a new drug or new indication, we'd recruit prime targets (who of course weren't supposed to know they were targets) to participate in "post-marketing studies." We'd ask each physician to enroll ten patients, take samples from us, and write prescriptions. We'd give them a "starter kit" with patient education materials and registration forms. And we'd pay them $250 to $500 per patient. If other methods didn't get them to write, this one would. After giving a drug to ten "study" patients for a while, a doctor would be pretty darn comfortable in writing prescriptions for regular patients. Whatever the number, after five, ten, fifteen, twenty patients, they'd start writing. The profit margins were so huge.

As the protagonist of *Side Effects*, Slattery-Moschkau cast TV actor Katherine Heigl, who easily meets Hollywood standards for physical attractiveness. In a film about a pharmaceutical rep, that is a faithful representation of reality. A 2004 paper in the *International Journal of Medical Marketing* sought to assist drug companies in their sales efforts by providing "a theoretical model illustrating the formation of physicians' perceptions of pharmaceutical sales representatives." The authors noted that:

> Empirical studies of gender and attractiveness, in the employment context, have generally supported the notion that attractiveness is beneficial to men; but varies for women, depending upon the nature of the job ... physical attractiveness work[s] against women applying for managerial positions, but [is] an asset for women seeking nonmanagerial positions. The sales position is typically perceived as being subservient to

the customer. Consequently, physical attractiveness, for both men and women, should have an equally positive impact on physician perceptions of the pharmaceutical sales representative.[39]

The companies didn't need experts to tell them that; young people with conventional good looks are the rule among drug reps. And that may become even more pervasive. The same article observed that as more restrictions are placed on rep–doctor relationships, it will no longer be possible for reps to "buy rapport." So, say the authors, "it will be more critical than ever for pharmaceutical companies who continue to use sales representatives, to focus on how these representatives are perceived in the physician's office."

Having been out of the drug business since 2002, Slattery-Moschkau has kept a close eye on it and sees only small improvements:

> I still have friends in the industry, believe it or not, and they say it's still pretty much the same. Recently, I was sitting in an airport lounge next to a very well-dressed, very nice-looking young man, a "Ken-type" [as in Barbie's boyfriend], and I instinctively wondered to myself if he was a rep. He ended up sitting across the aisle from me on the plane. Sure enough, he opened up a big drug-company binder, and I could see the heading on the first page: "Heavy Hitter List." It showed the top-prescribing doctors he was zeroing in on.

"Nothing has changed," concluded Slattery-Moschkau.

Clearly, the drug companies have no intention of calling off their quest to accelerate the growth of pill-popping. That means more trouble ahead, and not just for the patient. In recent years, the pharmaceutical industry has become so tightly integrated across the planet that a patient in Madison, Wisconsin is connected directly through doctors, pharmacists, and sales reps from companies based in New Jersey or North Carolina or Switzerland to ingredient suppliers in India and China, and through them, to workers and farmers in some of the world's most impoverished and polluted places. One such place is Patancheru, India.

3

SIDE EFFECTS MAY BE SEVERE

The faster the economic process goes, the faster the noxious waste accumulates. For the earth as a whole, there is no disposal process of waste. Baneful waste is there to stay unless we use some free energy to dispose of it in some way or another ... There is a vicious circle in using detergents for economy of resources and labor and afterwards having to use costly procedures to restore to normal life lakes and river banks.

Nicholas Georgescu-Roegen, *The Entropy Law and the Economic Process*[1]

Oxen working the fields, the eternal river Ganges, jeweled elephants on parade. Today these symbols of ancient India exist side by side with a new sight—modern industry ... Throughout the free world, Union Carbide has been actively engaged in building plants for the manufacture of chemicals, plastics, carbons, gases, and metals ... A hand in things to come.

From a Union Carbide Corporation newspaper advertisement, circa 1961, which featured a painting of a gigantic, pale hand pouring a chemical over an Indian farmer's field. Twenty-three years later, a gas leak from a Union Carbide factory in Bhopal, India killed tens of thousands in history's worst industrial accident.

The legal drug trade between India and Western nations is booming, but it's also in a constant state of flux, and—for anyone wanting to know who's selling what to whom—opaque. Switzerland-based KPMG International, "a global network of professional firms providing audit, tax, and advisory services," provided a rare peek at that trade in its 2006 report to investors entitled "The Indian Pharmaceutical Industry: Collaboration for Growth." The report says that India's drug exports are growing at a rate of 30 percent annually and that

Indian drug manufacturers currently export their products to more than 65 countries worldwide. Their largest customer is the U.S., the world's biggest pharmaceutical market. The use of generic drugs is growing quickly in the U.S. due to cost pressure by payers and the introduction on January 1 this year of the Medicare Part D prescription benefit, giving seniors and people with disabilities prescription drug coverage for the first time. With 74 facilities, India has the largest number of U.S. Food and Drug Administration (FDA)-approved drug manufacturing facilities outside the U.S. Indian firms now account for 35 percent of Drug Master File applications and one in four of all U.S. Abbreviated New Drug Application (ANDA) filings submitted to the FDA. Analysts at Credit Lyonnais Securities Asia say they expect the number of generic drug launches by Indian companies in the U.S. to increase from 93 in 2003 to over 250 by 2008.[2]

But export of generic drugs is only one aspect of India's key role in the world pharmaceutical business:

- India is one of the world's leading manufacturers of bulk drugs. Technically known as "active pharmaceutical ingredients"—raw materials for making pills, capsules, etc.—bulk drugs are important to India's domestic industry but are increasingly exported. India and China together supply as much as 20 percent of the US market for generic and over-the-counter drugs and 40 percent of all bulk drugs used here—and that may rise to 80 percent by 2022. India's share of the US market in 2006 was $800 million, exceeding China's.[3]

- Indian drug companies are signing stacks of contracts to supply bulk drugs and intermediate compounds to Western companies, including Pfizer, GlaxoSmithKline, Eli Lilly, AstraZeneca PLC, and Merck & Co. The many finished drugs made from those compounds include pantoprazole (Protonix) and esomeprazole (Nexium) for acid reflux; ranitidine (Zantac) and Nizatidine (Axid) for ulcers; levobunolol (Betagen) and brimonidine (Alphagan) for glaucoma; eprosartan mesylate (Teveten) and losartan (Cozaar) for high blood pressure; and methohexital (Brevital) for anaesthesia.[4]

- Pharma multinationals are selling again in India, since the 2005 India Patents Act brought tighter enforcement of intellectual property rights. KPMG tells Western investors that "New government initiatives seek to enable the majority of the population to access the life-saving drugs they need, while even greater opportunities may be presented by the rise of the new Indian consumer. This group—urban, middle class and wealthy—live fast-paced, Western-style lives and, as a result, they are beginning to suffer from Western, lifestyle-related illnesses, for which they want, and can afford, innovative drug treatments."

- In India, 100 percent foreign ownership is allowed in pharmaceuticals, while foreign investment continues to be limited to less than 50 percent in most industries.

- Indian pharmaceutical companies are buying up smaller firms in Europe and America at an accelerating pace.

- Companies find that India is not only a cheap place to make drugs but also a cheap place to test them. There is a large population of people living on low incomes for whom the small payments that go to drug-trial participants are a seeming windfall.

COLORFUL INDIA

As a result of all of these forces, reported India's Associated Chambers of Commerce and Industry in 2006, the nation's pharmaceutical industry was in the midst of a rapid growth spurt, estimated at 11 percent per year.[5] The Indian government is pushing for creation of "Special Economic Zones" across the nation for drug manufacturing, estimating that a single 125-acre zone can generate annual exports of almost $500 million.[6]

Even before the establishment of Special Economic Zones, India's pharmaceutical industry was concentrated in a few small areas, one of the most notorious being located in and around the town of Patancheru in the state of Andhra Pradesh. Over the past two decades, a growing chain of industrial estates has turned this stretch of countryside into a toxic hot spot. Numerous plants are turning out both finished and bulk drugs. Thirty-five to 40 percent of India's bulk-drug manufacturing is done there; that's a lot of production, and more waste than Patancheru can handle.

Competition in the international drug market is fierce, and corner-cutting on waste treatment is common. Industrial estates in and around Patancheru make routine drugs for the domestic market as well as a wide array of bulk drugs and intermediates that are used to make products of global interest. In addition to those listed above, they make ingredients for antibiotics, anti-HIV drugs, antidepressants, statins to lower cholesterol, cardiovascular drugs, and remedies for athlete's foot and osteoporosis. Some of these products, designed to alleviate diseases of affluence, would hold little interest for most of India's rural population. But the people who live and work in the area appear to be sicker than average, hit with diseases of the rich as well as those of the poor.

I visited Patancheru in 2005 and ventured into the surrounding countryside, which is drained by a now-dying stream called the Nakkavagu. The watershed, about 10 miles by 25 miles, was once good farmland, crisscrossed by intermittent streamlets, their water conserved in a chain of 14 small, picturesque reservoirs called *cherus* that provided irrigation water during the long dry season. But by the 1970s, the area's good water, wide open spaces and proximity to Hyderabad (India's fifth largest city) made it a prime target for economic development. It wasn't long until the reservoir for which Patancheru was named had become a stew of toxins. Since the late 1980s, a half-dozen large industrial estates have cropped up across the surrounding countryside. Over that time, the word "Patancheru" has acquired multiple meanings: a lake, a town, a region, and, in more recent years, an ecological catastrophe.

In travel books, India is inevitably described as "colorful." I found that to be true in the Patancheru area, but with a nasty twist. Where National Highway 9 crosses the Nakkavagu just west of town, its sluggish waters appeared deep purple. In a dump outside the Isnapur industrial estate south of the highway, an alien landscape of grey ash mounds was dotted by clumps of burnt-orange and sulfur-yellow powder. Asanikunta, a lake bordering the Bollaram industrial estate, had the color of cabernet sauvignon and the aroma of paint thinner.

Accompanied by some local residents, I passed through the Kazipally estate, home to an assortment of chemical and pharmaceutical companies, and stopped at a factory identified by a sign as belonging to SMS Pharmaceuticals. Behind the plant, tar-like liquid was dribbling over a concrete dam, running down a deep gully, and meandering through a barren field beyond. A second sign in front of the plant, erected by court order, listed some of the chemicals being used inside: toluene, methyl isothiocyanate,[7] DMSO, chloroform. It was nearly impossible to breathe anywhere within a hundred yards of the plant, and hard not to retch.

Dr. Allani Kishan Rao has practiced medicine in Patancheru for 30 years, and he's been fighting pollution for most of that time. He told me in 2005, "Illness rates here are more than 25 percent, compared with 10 percent nationally. I'm sure it is related to the chemical intermediates, organic solvents, and gases that come out of the industrial estates." Finished drugs undergo extensive safety testing, but, said Dr. Rao, no one knows what the smorgasbord of molecules currently flowing from the drug plants is doing to his patients, especially when combined with pesticides and heavy metals from other factories. He's intent on seeing his own country's laws better enforced. But now that affluent patients everywhere are directly connected to Patancheru via the global pharmaceutical market, he also has a message for Western consumers: Our hunger for drugs, whetted by industry's hunger for profit, is helping keep his own patients sick and broke.[8]

According to *Washington Post* reporter Marc Kaufman, the FDA conducted 1,222 quality-assurance inspections of US drug-manufacturing plants in 2006. That same year, the agency carried out only 32 inspections of Indian drug plants, mostly to inspect companies applying to export to the US, not to check on the quality of products from existing suppliers. And, reported Kaufman, "on-the-ground inspections of Indian and Chinese plants remain rare and relatively brief and are always scheduled in advance, unlike the surprise visits that FDA inspectors pay to domestic manufacturers."[9] And there is no indication that FDA inspectors pay any attention to environmental impacts of the plants. So whatever the quality of the exported drugs, by the time they're swallowed in Chicago or Atlanta, the consequences of their manufacture might already be felt by Indian villagers downstream or downwind from the drug factories that made them. Those side effects are never mentioned on the pill bottle or in the pharmaceutical companies' direct-to-consumer advertisements.

SIDE EFFECTS

In October 2004, Greenpeace India released the results of a study comparing the health of almost 9,000 people in nine Nakkavagu-basin villages with a control group of four villages in a non-industrial part of the same district. Cancer rates in the nine affected villages were 11 times as high as in other villages. Rates were 16 times as high for heart disease; four times as high for birth defects; and two to three times as high for skin problems and disorders of the nervous, endocrine, and metabolic systems.[10] Dr. Rao said that the Greenpeace results were consistent with his clinical experience: "I am seeing far more cancers, heart ailments, birth defects, and epilepsy than I did in the '70s and early '80s, especially among children. And I'm seeing a lot of tuberculosis, suggesting that immune systems are being compromised, too."

High disease incidence in the area should come as no shock. According to the 2004 report of a "fact-finding committee" appointed by the state High Court of Andhra Pradesh, pollutant concentrations in area streams and lakes range from 12 to 100 times as high as those in an unpolluted lake just outside the contaminated zone.[11] Partly as a result, and in accordance with court orders, drug companies are paying to have safe water piped into affected villages for drinking and cooking. But the polluted water is still used for other purposes in the home and on the farm. Walking alongside a stream near the village of Gandigudem, I came across a woman doing her laundry. As she slapped a shirt against a rock, she told me through a local translator that her name was Nagamani. I asked if she ever carried any of the stream-water home. "Oh, no, I only wash clothes here," she said. "If you bathe in it, you get a rash." In the nearby village, a farmer confirmed Nagamani's diagnosis. Opening his shirt, he pointed to a broken-out area on his neck and chest. And, he said, "I walk a few steps, and I'm out of breath. And I'm only 39 years old!"

Thousands of acres of formerly good farmland around Patancheru lie uncultivated during the dry season because groundwater has become unfit for irrigation. The High Court fact-finding committee sampled 48 wells in the area and found 81 percent of them polluted beyond an international standard for irrigation water.[12] In Gandigudem, a farmer named Janardhan showed me his abandoned plow, his junked rice mill and his idle irrigation well and pump. "Around here, if we have good water we can survive," he said. "Now, without good water, we're finished." He pointed to a plot of a few acres where he once grew rice. A group of boys were using the abandoned paddy as a cricket field. A lone water buffalo grazed on some weeds. Asked what becomes of the buffalo's milk, Janardhan said, "We'll sell it at the market, but we sure won't drink it."

A neighbor, Rekha, told me her subsidized rice ration never lasts till the end of the month. "Before the pollution, we grew our own rice. Now I can't grow crops, four of my water buffalo have died, and we're forced to depend on the government." Another ruined farmer, Rajaiah, showed me photographs of three water buffalo for which he had paid 20,000 rupees (almost $500) each. The animals, to all appearances well-fed, lay dead in a patch of tall grass where they had collapsed a few months ago. A half-born calf protruded from one of the cows.

The analysts at KPMG estimate, "The 'organized' sector of India's pharmaceutical industry consists of 250 to 300 companies, which account for 70 percent of products on the market, with the top 10 firms representing 30 percent. However, the total sector is estimated at nearly 20,000 businesses, some of which are extremely small."[13] The two types of companies are components of a single system, with the small ones often feeding intermediate compounds and other chemicals to the larger ones. Both are well-represented in Patancheru's industrial estates.

At least superficially, conditions at some of Patancheru's major drug operations don't seem too bad. One of the country's ten largest companies, Aurobindo Pharma, makes generic and bulk drugs for antibiotic, antiretroviral, cardiovascular, central nervous system, gastroenterological, and anti-allergy treatment. As of 2006, the firm had filed 324 applications with FDA or European drug agencies.[14] In January 2005, I stopped outside Aurobindo's Unit Number 5 in the Isnapur estate. There's a strip of green grass around the factory perimeter, and the company has planted a mini-forest of eucalyptus next to a small shantytown across the road. There I watched residents of the huts line up with water pots outside one corner of the factory, where a hose providing safe, trucked-in drinking water had been flung over the wall for their use. None of the Patancheru companies, including Aurobindo, would respond to my requests to discuss the situation. But later that month, I took a guided tour of a plant lying 50 kilometers to the west, far outside the Patancheru valley. The managers of that factory, which is owned by another top-ten company, Nicholas Piramal India Ltd. (NPIL), were taking effective steps to limit pollution, at least to my untrained eye. Nevertheless, residents of a nearby village complain bitterly that their groundwater has been polluted by the plant, and NPIL, under a court order, was paying to have clean water piped in for them.[15]

Back in Patancheru, Dr. Rao agreed that some improvements have gone beyond the merely cosmetic. More companies, for example, are starting to install at least minimal pollution control systems. But in many cases the more toxic production steps have simply been outsourced: "Go to the main units of the bigger companies, and it will all look very green," he says. "Just be careful not to go to places like Kazipally from where they get their intermediates and solvents."

The High Court committee visited 40 companies with "high pollution potential" in the industrial estates. Of those, 30 were producing drugs or drug ingredients, and only five of those were complying fully with Patancheru's already too-lenient pollution laws. For effluent at new US drug plants, the Environmental Protection Agency sets strict limits on at least 34 chemical compounds, from acetone to xylene. But in the Patancheru area, where normally only the total quantity of pollutants is tracked, there's almost no information about specific toxic compounds. That's serious, because some of the drug industry's solvents, byproducts, and ingredients can harm people even at low concentrations.

Since the mid 1990s, all area factories have been legally required to haul their toxic wastes to the Common Effluent Treatment Plant just outside Patancheru, on a tributary of the Nakkavagu. A local resident led me down a ravine to see the pipe through which the plant releases treated effluents back onto the landscape. Pouring from the pipe was a thick stream of coffee-colored liquid topped with thick, white foam that has often been found to carry pollutants at many times the statutory limits.[16] A recent Swedish study found astonishingly high concentrations of eleven drugs – antibiotics and treatments for high blood pressure, ulcers, and allergies – in this devil's cappuccino. The research team noted that "to the best of our knowledge, the concentrations of these 11 drugs were all above the previously highest values [ever] reported in any sewage effluent."

CRACKING DOWN?

Venkat Ram co-owns Denisco Chemicals, a Hyderabad company that produces small quantities of specialized compounds (called "fine chemicals" in the trade) and has a correspondingly small flow of wastes that he can handle properly at relatively low expense. He told me, "I would not like to be in the position of the big corporations that manufacture bulk drugs. They have enormous production runs, as compared with my small volumes, and their waste-disposal costs are correspondingly huge. That's a serious problem in their market, with the cutthroat price competition."

But, said Ram, "The drugmakers I know are very careful to abide by environmental regulations. To survive these days, they have to export into regulated markets, and that means they have to follow strict international standards set by the World Health Organization." He feels that India's burgeoning exports can't be blamed for pollution of the Nakkavagu basin; on the contrary, he said, international markets are creating pressure for cleaner, safer manufacturing practices. "It's not like the '80s and '90s when guys were dumping all over the place."[17]

Companies may not be as reckless as they used to be, but the sheer volume of production is causing the system to overflow. Official responsibility for cracking down on the state's polluters falls to the Andhra Pradesh Pollution Control Board (APPCB), but critics claim the state Board is not enforcing the law in Patancheru. Back in 2003, India's national Supreme Court in New Delhi assigned a committee to monitor the enforcement of toxic-pollution laws in Andhra Pradesh and four other states. Together, the five states account for 80 percent of the nation's hazardous waste problem, and Patancheru is one of the localities on which the Supreme Court Monitoring Committee (SCMC) has focused. In a February, 2005 letter to state officials, the SCMC complained that "the entire efforts of APPCB to catch defaulters appear to be only rudimentary and lacking earnestness."[18] The clearly exasperated SCMC further noted a large and dangerous accumulation of hazardous wastes at Patancheru's Central Effluent Treatment Plant, despite which the plant "is merrily continuing its operation in absence of any stern actions taken by APPCB." By March 2006, more than a year after my first visit to Patancheru, members of the SCMC themselves appeared to have become potential customers for the companies' mood-lifting drugs; their report issued at that time noted that "SCMC is quite depressed with the physical condition of several of the industrial estates in Andhra Pradesh."[19]

I returned to India in 2007 to find that little progress had been made in cleaning up the Nakkavagu basin. I spoke with meteorologist Dr. S. Jeevananda Reddy, who once traveled the world as a chief technical advisor to the United Nations but is now back in his home city of Hyderabad pushing for tougher policies on pollution. From 2002 to 2006, he was one of three outsiders serving on an APPCB task force assigned to monitor pollution in the Patancheru area. But Reddy told me that the companies and their friends in the government call the shots. Many of the problems uncovered by his task force went unaddressed, he said, because the APPCB was created by the state legislative assembly and "has to oblige the politicians, who are always telling them, 'So-and-so should not be punished'." He says one pollution-control official who managed to halt the flow of toxic water from an SMS Pharmaceuticals plant (the same noxious facility I'd seen in 2005) was not congratulated, let alone rewarded, by APPCB; he was suspended for being too gung-ho.

"In the USA," Dr. Reddy told me (perhaps a little naively) "any company doing mischief will be stopped from selling its product. Here, people can't even find out which ones are polluting. We asked the APPCB simply to list violators on the Internet. They wouldn't do it." He says the biggest problem is the sheer quantity of drugs that plants are producing, which means that they pump out far more waste water than the treatment plant can handle. The state permits each company to dispose of only a certain amount of water per day, and if its chemical concentration is too high, the company is fined. But, said Dr. Reddy, "The fines are peanuts to them." He claimed that factories stay within emission limits by "giving their highly polluting processes to outside job-workers who have no treatment capacity."[20]

Some companies move part of their drug production under their "research and development" wing to sidestep limits, but even that doesn't take care of it all. Aurobindo, Dr. Reddy told me, was at one point running at ten times the permitted output. "To shut them down, the state electricity board cut off their power, but after political intervention, the lights were back on in a few days and they were back in production." He himself has seen what often happens when tanker trucks are turned away by the effluent treatment plant once its daily capacity has been exceeded: they dump it in the countryside before returning to the plant.

The Intergovernmental Panel on Climate Change, recognized as the world authority on the subject, has made it clear that the same wealthy nations that created the global-warming problem will be able to buy more protection against it than can poor nations.[21] Nevertheless, Dr. Reddy attributed rising ecological awareness in Western nations to the fact that rich nations, unaccustomed to the kinds of environmental nightmares that routinely afflict the poor of India, will actually feel the effects of global climate change:

> Now, all they want to talk about in the West is warming, warming, warming, because they will feel it. But we cannot ignore the direct impact that toxic industries are having on humans, animals, and plants in places like Patancheru. The tragic results are right there in front of us. Doing something about that will help correct the global problems!

I returned to Patancheru to visit Dr. Allani Kishan Rao, two years after we had last spoken. He had very little change to report: "All the government has done is to make Patancheru a Special Economic Zone. It was specifically chosen to host more drugmaking factories because it's already so polluted. So they're doing a lot of road-widening but nothing to improve the soil and water."[22] The "Special" designation has also driven up land prices and, believes Dr. Rao, it will eventually chase agriculture out of the area altogether. It appears that the state's plan to deal with toxic wastes from export-led growth is simply to concentrate it in the basin of the Nakkavagu—now an official ecological sacrifice zone.

As in other areas of trade, the relationship between the US and India in pharmaceuticals is not symmetrical. The US market's huge growth and the prospect of even greater growth have done much to drive the greatly increased output of drugs and intermediates in India, which now has 75 FDA-approved pharmaceutical plants, more than any other country outside the US. One projection has India's output going from its current $9 billion per year (more than a third of that being exported) up to $25 billion in 2010.[23]

The costs of this growth to human and ecological health are not being tallied. (A 2004–5 study by a major Indian university was aimed at determining the health impacts of pollution in the area, but when government officials previewed the results, they apparently didn't like what they saw; they told the researchers to hold off on publishing, go back, and gather more data.[24]) Sudden calamities like Union Carbide's 1984 poison-gas leak that killed tens of thousands in Bhopal, India generate closely watched casualty counts. No comparable statistics exist for Patancheru's long-running tragedy, but Dr. Rao believes the toll will climb for years to come. He told me, "We're now into a second generation of toxic exposure. We're seeing another Bhopal, but in slow motion."

An American population that maintained better general health would be a much poorer source of profits for the pharmaceutical industry and would put a lot less pressure on places like Patancheru. One of the most widely recognized sources of ill health today is obesity; based on current definitions, the government says almost two-thirds of Americans are overweight or obese. That has opened up great opportunities for a long line of drugs and other profitable weight-loss remedies over the years. But the most popular "cures" for overconsumption of food tend to result in more, not less, resource consumption.

4

SWALLOWING THE EARTH WHOLE

In late 2003, with the tide of anti-carbohydrate mania rising fast, I asked my colleague Marty Bender for advice. Marty was an expert analyst of energy and resource consumption, and I wanted his help in answering an entirely hypothetical question: What would be the ecological consequences if everyone on earth who wanted or needed to lose weight adopted the Atkins diet?[1]

Atkins Nutritionals, Inc.—at that time the heavyweight champion of the low-carb world—has since gone bankrupt and then emerged from bankruptcy. But back then, the regimen popularized by the late Dr. Robert Atkins, which featured lots of protein, ample fats, and tightly restricted amounts of starches and sugars, was boosting national and global sales of meat, fish, and eggs (as well as more exotic, carbohydrate-free, high-fat foods like pork rinds). Meanwhile, it was putting the squeeze on makers of orange juice, bread, rice, and other high-carb staples of the American diet.

Marty and I obviously didn't really expect that every overweight[2] person on earth would join the Atkins rush. But we knew that it takes more resources to produce proteins and fats than to make sugars and starches, especially when food is produced as it is in the industrial West. That means that, other things being equal, an Atkins dieter would probably leave a heavier ecological footprint than would a conventionally omnivorous eater or a vegetarian. It's reasonable to ask, when faced with a strong societal trend like low-carb eating, "What if everyone did it?" When such a question is answered in ecological terms, it helps distinguish globally unsustainable practices from others that might come closer to being sustainable. And, of course, it highlights fairness. You may recognize this approach as bearing some resemblance to eighteenth-century German philosopher Immanuel Kant's proposition that if you're a rational, moral person, you will "act only according to that maxim by which you can at the same time will that it should become a universal law."[3] Our purpose, however, was not to prescribe or proscribe individual behaviors but to evaluate the consequences of mass behavior. Even to ask the question is to question capitalism itself, because to rule out any type of luxury consumption would be to wall off vast, profitable opportunities.

For decades, business has been coming up with "solutions" to the problems that result from America's overconsumption of food and underexertion of bodies; however, as with the Atkins diet and other trends we will consider in this chapter, the answers almost always require even more, not less, consumption. We are left to wonder, "Should people leading Western lifestyles remedy their own overindulgence by indulging in even more consumption—more than the majority of the world's people could ever experience?" But to ask the invisible hand of the market to handle that question would be like asking the Oracle of Delphi for directions to Disney World.

CAUTION: THIS DIET IS NOT FOR EVERYONE

Even before doing our calculations, Marty and I knew that making a high-protein, high-fat diet universal would increase the burden on the planet. It takes 68 times as much water to produce a pound of beef as it does to produce a pound of bread flour, and animal protein is eight times costlier than plant protein in terms of fossil-fuel energy.[4] The industrialized West has managed nonetheless to put meat and dairy products within reach of even its poorest citizens, at a high cost in soil, water, and energy.

To determine just how hefty the global low-carb burden would be, we started with the Worldwatch Institute's estimate that one billion humans are overweight.[5] We then calculated that extending the Atkins diet to all of them would require the world's meat, dairy, poultry, and seafood industries to increase their output by 25 percent. With current methods of raising livestock, that would mean sowing almost 250 million more acres of corn, soybeans and other feed grains.[6] So much more land would be needed because feeding grain to animals and then eating their meat, milk, and eggs is much less efficient than eating plant products directly. Finding a quarter-billion acres for adequate feed-grain harvests would mean putting 7 percent more land worldwide under the plow—probably more, because farmers are already using the most productive lands. Much of the newly cropped acreage would likely be marginal, prone to greater erosion and in need of extra-generous applications of fertilizers and pesticides.

The resulting population explosion of edible animals would worsen air and water pollution and threaten even more workers with injury and misery, given the intensification of meat production that we'll see in the next chapter. Grain-fed cattle usually spend part of their lives grazing. New pasture for, say, half of the additional animals called for by a global Atkins diet would require another billion acres. Most of these new grasslands probably would come by deforestation, which could mean chopping down 10 percent of Earth's remaining forests. On the other hand, to give landscapes, livestock, and people a break and squeeze more protein from the already overfished oceans could be even more damaging.[7]

We had a simple reason for assuming that Atkins-diet animals would be raised in the same ecologically destructive way that current ones are: The greatly increased demand for protein would be met by the same industrial machine that created the obesity problem and conceived the low-carb solution. Were this a different economy, were the production system designed more to feed people better than it does balance sheets, moderate beef consumption could actually improve the environment. Cattle and other ruminants like sheep and goats can subsist entirely on well-managed range, pasture and hay. Extensive research shows that rangeland and pasture, which consist almost entirely of perennial plants with large, long-lived root systems, do not cause the devastating soil erosion and water contamination seen with annual grain crops.[8] North America's lands and rivers would be dramatically improved were we to take the 60 percent of grain acreage that feeds cattle and convert it to pasture. According to other of Marty's calculations, current production of beef and dairy products might possibly be maintained with such a system. But without massive deforestation, grass-fed beef still couldn't satisfy a billion people on a high-protein diet.

APPEARANCES AND REALITY

Binghamton University professor Bat-Ami Bar On has taken a look behind the remarkable ability of products like the Atkins diet to sell, despite the fact that they often are not "accommodating" (that is, they often don't fulfill one of the basic requirements of a commodity: to function properly). She observed:

> [I]t is important for the capitalists that commodities be perceived as accommodating whether they actually are or not ... Capitalists therefore have a vested interest in consumers taking surface appearances as reality, in a lack of distinction between appearances and reality, and in the creation of the most solid possible impression that the products that their enterprises put on the market are indeed real commodities. To a great extent this is what advertising is all about.[9]

Bar On went on to ask, "Is the Atkins Diet a commodity or is it a simulation or a conjured-up imitation of a commodity?" Now that Atkins has taken its bows and yielded the stage to competitors, we can look back and see that it didn't really matter whether it was a real or fake commodity. It did what commodities do, generating a lot of economic activity and using up a lot of resources. Despite scares over mad cow disease in 2004, the price of a live steer in Texas hit 84 cents per pound, up from 67 cents in 2002; turkey consumption in the state shot up 22 percent.[10] Delighted feedlot and poultry companies credited low-carb eating for much of the boost. The hog population of Iowa rose by 640,000.[11] Earnings on shares of Smithfield Foods, Inc., at the time the nation's largest producer of hogs, fresh pork, and processed meats, increased more than tenfold.[12] A couple of years later, the CEO of Tyson Foods, Inc., America's biggest meat producer, responded with a kind of nostalgia to a March 2007 report in the *Journal of the American Medical Association*[13] showing that the Atkins diet probably had achieved greater weight loss than some other popular diets: "Atkins was good for demand then and its accolades here recently ... that's good for us and we do appreciate that the Atkins diet gets that kind of recognition."[14]

Atkins was just one phase, if an especially newsworthy one, in the long, meandering evolution of Western dieting. In the years since Marty and I did our analysis, traditional low-carb regimens in the Atkins vein have largely given way to diet plans more focused on the "glycemic index" of carbohydrates and foods. Diet plans, diet foods, devices, drugs, treatments, organizations, and facilities come and go, but one thing never changes: the weight-reduction market never loses economic weight. In the US alone, sales were $58 billion in 2007, and Marketdata Enterprises, Inc. predicted that they would reach $69 billion by 2010. With the low-carb boom fading, Marketdata saw continued growth in diet plans, diet-food home delivery, diet pharmaceuticals, and bariatric surgery (which drastically reduces the capacity of the stomach).[15]

Marty and I analyzed the ecological impact of the Atkins diet in only one dimension: the nutrient composition of the food consumed. To my knowledge, the *total* environmental burden has not been estimated for Atkins or any other weight-loss strategy. The foods and other commodities they offer are generally heavily processed, with high packaging-to-product ratios. Anything having to do with medicine can be ecologically pricey, as we saw in Chapters 1 to 3, and health clubs and weight-loss centers have an impact as well. All weight-loss products and services create a bigger burden than do those old-fashioned, well-proven measures that will be recommended by any nutritionist who isn't trying to sell you something: eating less, eating out rarely, cooking with food in its least-processed form, limiting consumption of animal products, drinking mainly water, avoiding between-meal snacks, and, whenever possible, walking, running or cycling instead of driving. To have all overweight people follow that and other prosaic advice for good health would avert conflict between humans and other animals; it would emphasize our reliance on natural systems; it would be more affordable for everyone regardless of income; and it would probably precipitate an economic crisis.

In a hungry world like this one, to be able to adopt any formal dietary/fitness regimen for purposes of self-improvement is a luxury in itself. In his novel *The Comedians*, Graham Greene has the narrator, Mr. Brown, make that point to the altruistic Mr. Smith regarding Smith's failed proposal for a "vegetarian center" in Haiti:

> Brown: "I don't think they are quite ripe here for vegetarianism."
> Smith: "I was thinking the same, but perhaps..."
> Brown: "Perhaps you must have enough cash to be carnivorous first."[16]

Nutrition schemes make excellent commodities because they are perennially popular whatever their failure rate. (And failure is the norm; according to a National Institutes of Health panel, "In controlled settings, participants who remain in weight loss programs usually lose approximately 10% of their weight. However, one third to two thirds of the weight is regained within 1 year, and almost all is regained within 5 years.")[17] If they did their job well, or if people swore off them for good whenever they failed, the market would slow to a trickle. But the success of weight-loss plans isn't entirely the result of failure; their tag-team partner, the food industry, ensures a steady flow of lapsed dieters seeking a second or a fourth chance. Newly overweight customers seem to grow younger every year. It takes the ideas of entrepreneurs, an excess supply of fattening foods, and plenty of sedentary jobs and couch-potato pastimes to keep the weight-loss game going.

And it's finding new frontiers all the time. Advertisements for weight-loss plans, diet foods, and fitness centers are now ubiquitous in the cities of India, a country where 25 percent of upper-income women[18] and 30 percent of upper-income adolescents[19] are now clinically obese even while 21 percent of urban women and 48 percent of rural women are undernourished.[20] It's no more than one would expect in India, a country where 20 percent of the population eats 80 percent of the dietary fat.[21] There, capitalism has been embraced with an ardor rarely seen even in the West; the front half of every Indian bookstore I've visited is stuffed with business, management, and motivational books.

The more wealth there is in a given region or social stratum, the bigger the share of commodities it can absorb. With obesity, capitalist economies are well-tuned for supplying commodities that create the problem—rich and plentiful food, motor vehicles, TVs, computers, video games—as well as those billed as solutions—diet books, diet foods, gyms, and drugs—but only to those who can afford them. Meanwhile, members of India's impoverished majority remain pleasingly lean (if they're managing to obtain a sufficient diet) or emaciated (if they're not).

America and other Western nations—and even some poorer ones[22]—have broken out of the historical pattern that says the rich shall be fat and the poor thin. Here, food, especially low-nutritional-value, fattening food, is plentiful and cheap while commodities advertised as "solving" the problem of excessive weight gain are not. In 2006, *Forbes* magazine surveyed the costs of ten of the most popular weight-loss plans. Not surprisingly, all of them added significantly to the dieter's weekly food bill, with the median increase pegged at 58 percent. The Jenny Craig plan was the most expensive, boosting food costs by 152 percent. Atkins was number three, with an 85 percent increase. Nutrisystem added 109 percent, Weight Watchers 78 percent. The strategy of eating low-fat sandwiches at Subway restaurants, heavily publicized on TV, was the cheapest, adding 26 percent.[23]

But even Subway is a luxury. In the 2004 documentary *Supersize Me*, filmmaker Morgan Spurlock ate nothing but McDonald's food for a whole month. The most poignant scene in the film occurred on a side-trip to visit a competitor. At a Subway restaurant, Spurlock featured an eighth-grader named Victoria who had come with her mother to an event featuring company spokesperson Jared Fogle, who is widely celebrated for having lost more than 200 pounds on a Subway-based diet. Fogle told Victoria, who was immersed in her own struggle to lose weight, "The world's not going to change—you've got to change." When Fogle had moved on to other customers, Victoria spoke to the camera: "I guess it's kinda cool that I know somebody and can be able to listen to somebody about actually being where I am now, and it's hard because I can't afford to go there like every single day and buy a sandwich, like, two times a day, and that's what he's talking about, like that's the only solution."

ALL THE FISH IN THE SEA

More and more, nutritional plans emphasize not food but the individual compounds into which food can be broken, like sugars, starches, proteins, fatty acids, fiber, vitamins, and anti-oxidants. In the food game, the sum of the parts turns out to be worth far more than the whole, so the decomposition of food has opened up whole new marketing vistas. Now food itself is losing its definition, as what we eat is increasingly regarded as a simple agglomeration of nutrients to be consumed in proportions prescribed largely by the sellers of the nutrients. Much of this has been prompted by a real problem: the distorted diets that have evolved in an increasingly urbanized world. Whole industries have emerged to plug the holes that industrial agriculture and food processing leave in the human diet.

In examining an especially critical need to plug one such hole, British writer George Monbiot has drawn attention to a problem that shows how difficult it can sometimes be to answer Marty's and my pragmatic—or Kant's philosophical—question, "What if everyone did it?" Noting that before the dawn of agriculture, people consumed approximately equal amounts of omega-3 and omega-6 fatty acids (two of the many kinds of fatty acids that, strung together, make up fats and oils), Monbiot cited figures showing that we in the West get only one-seventeenth as much omega-3 as omega-6. He then discussed studies suggesting that childhood neurological problems like dyslexia and ADD are associated with a deficiency of omega-3 fatty acids, especially in the womb.[24]

As is well known and as Monbiot pointed out, the highest concentrations by far of omega-3s, and the highest omega-3/omega-6 ratios, are found in oily fish species. Incorporated into convenient capsules, the omega-3 fatty acids from fish might make a big difference in kids' school performance, not to mention their lives in general. Treating millions of children worldwide would be a relatively simple and effective procedure. But, as Monbiot put it, "There is only one problem: there are not enough fish." Citing Charles Clover's 2006 book *The End of the Line: How Over-fishing is Changing the World and What We Eat*[25] Monbiot went on to list some of the many ways the human economy, using factory-style vessels and methods, has plundered and then frittered away its nutritionally priceless ocean catch, using it as fertilizer, fuel, and feed—to nourish terrestrial livestock and even other fish. (I would add that a third of all canned fish sold in the US is for cats and dogs.[26])

A study of ocean ecosystems published in the journal *Science* made headlines in November 2006 by predicting the global collapse of all currently exploited fish species by the year 2048.[27] A few other researchers have quarreled with the predicted date of collapse, but few argue that exploitation of the oceans can continue at its current level for the long haul. Any attempt to reverse the growth of fishing will have to fight its way upstream against continuous growth in demand, as persistent low-carb publicity has thrown its weight behind the traditional reputation of fish as generally healthful. In recent years, medical studies have indicated many benefits of fish-eating beyond those cited by Monbiot. They have suggested, for example, that increased intake of omega-3s can alleviate heart disease and arthritis and curb the symptoms of bipolar disorder, schizophrenia, and Huntington's disease.[28]

Fish are rich in the two omega-3s of greatest interest, eicosapentaenoic acid (EPA) and docosahexaenoic acid (DHA), but they're not the only source. An array of plant-derived foods, flax seed being the most prominent, are sources of omega-3s that the human body can break down, albeit inefficiently, to EPA and DHA. Some vegetable oils have much higher omega-3/omega-6 ratios than do others, a quality that is considered nutritionally important. And a big shift away from factory farming could improve meat-eaters' health without any change in their daily menu. That's because meat and dairy products from grass-fed livestock and eggs from free-range chickens have much better omega-3/omega-6 ratios than do products from their feedlot- or confinement-raised counterparts.[29]

Using knowledge like that and moving wholesale toward nutritionally and ecologically balanced food systems would go a long way toward resolving the fatty-acid imbalances that are affecting human health. To ask, "What if everybody did *that*?" does not raise any obvious ecological dilemmas, as long as a sufficient quantity and variety of food is produced. But that approach has spurred only minor interest. On the other hand, treatment of the fatty-acid problem as an isolated medical condition is a "solution" with the kinds of qualities that spell success in a capitalist economy: an easily identifiable product (in this case, either a plate of fish or an oil capsule), a richly concentrated source of the product's essential ingredient that's easily mined (at least until it runs out); a simple marketing message (that the product is essential to ward off specific diseases, especially children's diseases); and an already well-established, and profitable, marketing context (the perceived need for many different nutritional supplements, each to solve a different problem). Isolating and treating fatty-acid imbalance indeed provides a lot of advantages to marketers but, as Monbiot implies, it would mean disaster if everybody did it.

The trend toward selling food part-by-part leads to countless other questions about how commodities could be made universal. Perhaps not every food constituent or supplement would create as much eco-havoc as can be projected for protein or omega-3 fatty acids. But to go through the list of all fatty acids, amino acids, sugars, vitamins, minerals, enzymes, antioxidants, and other compounds that have been, or are, or will be marketed as essential to good health, and to ask of each, "What if everybody did it?" would soon lead to at least one conclusion: For *everybody* to consume *a lot* of them, it's going to cost the Earth dearly.

Let's set aside products like a pack of Marlboros or a cold six-pack of Coors Lite or a Snickers bar or a Krispy Kreme donut or other traditionally guilt-inspiring commodities. Rather, in considering the effects of universal consumption, let's pick just four commodities that have been sold explicitly as means of achieving a longer, healthier, happier life: green tea, shark cartilage, Hoodia, and water.

Green tea. In the early 2000s, green tea reached the top of the antioxidant heap, promoted breathlessly as a preventive for cancer and cardiovascular disease and booster of the immune system. The FDA disagreed, announcing in May, 2006 that "FDA concludes there is no credible evidence to support qualified health claims for green tea or green tea extract and a reduction of a number of risk factors associated with CVD [cardiovascular disease]."[30] The Tea Council of the USA limply responded: "We anticipate that the research will evolve to support a health claim in this area in the future, since the anecdotal evidence certainly supports this position."[31] By that time, the number of "tea cafes" in the US had grown from 200 to 2,000 in ten years, and US tea sales were expected to rise from $6 billion in 2005 to $10 billion in 2010.[32] And the market still saw plenty of wide-open spaces yet to be conquered. If the extravagant claims made for green tea were eventually supported by research, there would be no reason for *anyone* not to drink it daily.

There appears to be no limit to what tea can be credited with. A website called The Tea Treasury exists "to enlighten people to embrace the tea drinking experience as a way of life"—that is, to sell more tea. The site's operator implies plenty while claiming little: "As a survivor of breast cancer, hepatitis C and a chronic heart condition, I have found tea drinking to be very beneficial to my health."[33] Tea is being marketed as a healer of the soul as well. In a 2007 *Utne Reader* article, Andy Isaacson chronicled the marketing of tea as a long-overdue, romantic antidote to coffee, America's dark beverage of "speed and productivity." The co-founder of Numi Tea told him:

> An impetus for me in designing Numi's packaging was to infuse the most mundane activity, walking through a grocery store, with the sublime and [a feeling] rich in self-reflection. What other way to subversively create a revolution—in this sense, a spiritual revolution—than through art and tea? I believe that tea, and art, are part of a feminine energy that is starting to permeate through our shift in consciousness.[34]

Isaacson identified the customer base being targeted by such flummery: "The sociologist Paul Ray calls these people 'cultural creatives' ... Cultural creatives care about ecological sustainability, social justice, and self-actualization." If cultural creatives are all for ecological sustainability, that feeling may run head-on into their craving for tea. I know of no quantitative analysis, like the one Marty and I did for the Atkins diet, that evaluates the impact of universalized tea-slurping, but it certainly would require landowners and corporations to expand the area under cultivation immensely. Responding to good prices, more of the highland tropics would soon be covered in picturesque tea plantations. But being deep green in color doesn't always make a scene ecologically green. Every tea-producing acre sits where once sat an ecosystem, usually a forest or high grassland, that had come, through evolution and co-evolution of its myriad component species, to be thoroughly adapted to its locality. Along with the precipitous loss of biodiversity that comes with conversion of a landscape to tea production, there is typically an increase in soil erosion, disruption of streams and rivers on the slopes and in the valleys below, depletion of soil nutrients, and a good dousing with pesticides and other synthetic chemicals.[35]

In January 2005, I visited Gurukula Botanical Sanctuary on the slopes of India's southwestern mountains, in the state of Kerala. Its director, Suprabha Seshan, showed me how the small non-profit is preserving a huge range of rainforest plant species threatened by deforestation, much of which has been done to open up land for tea plantations. In addition to preserving genetic diversity, Gurukula is gradually acquiring adjacent tea-producing land and allowing it to revert to forest. A year later, Seshan won the UK's top conservation honor, the Whitley Award, for her and Gurukula's efforts. More recently, I asked her about the potential for organic tea, which presumably would be more ecologically benign than industrial tea. Seshan told me that in regions where natural fertility is already in short supply, things are never simple:

> Organic tea is indeed a tricky issue. To support an acre of tea requires a lot of input and even though there is some attempt to use tea waste (from processing) and also the new coir [coconut husk] compost, there is still lack of supply, and this tends to mean that biomass gets lopped from vegetation bordering tea plantations and this unfortunately means deforestation.

If markets for tea, and especially organic tea, are to grow further, requiring expanded acreage, it could mean double trouble. As Seshan put it:

> If coffee farms get converted then that has a huge ecological impact because a multi-layered cropping system with lot of shade gets converted to shrub level plantation and this heats up and dries out a place that should be very cool and moist and diverse. The new plantations will come in areas that support taller vegetation—40 meters!—and if this gets lowered then this can be disastrous, first of all for watersheds. There is a lot of hype around organic tea and this is questioned by very few people.[36]

Expanded plantations would also require more exploitation of pickers and other tea workers, whose role, despite appearances, is not simply to be part of the scenery in those splendid mountainside tea gardens. Their day-to-day plight is not something most of us would want to experience; one of the hazards they face, ironically, is malnutrition. A study of workers on six tea estates in the Indian state of West Bengal found 41 percent of them with body mass indexes (BMIs) below 18.5; a population with more than 40 percent of its adults having BMIs below 18.5 is generally regarded as courting starvation.[37]

Shark cartilage. A 2005 study estimated that the numbers of sharks belonging to a key group of species in the Gulf of Mexico had dwindled to 1 percent of the population that had inhabited the same waters 50 years earlier. In the North Atlantic, shark populations had shrunk to 2 percent of their former size over the same period.[38] In 2007, a paper in the journal *Science* reported that the decline of large sharks has had "cascading" ecological effects, which have included the decimation of shellfish populations by the smaller sharks and rays that the big sharks used to eat.[39] As the authors noted in typically dry academic language, "Our study provides evidence for an oceanic ecosystem transformation that is most parsimoniously explained by the functional elimination of apex predators, the great sharks."

Many have pointed fingers for at least part of the blame at economic exploitation of sharks, including the widespread practice of "finning," in which the animals are discarded, sometimes dead, sometimes still alive, after their fins are harvested to be used in reputedly healthful gourmet soups.[40] Shark-fin soup is eaten primarily in East Asia, but sells for as much as $60 a bowl in San Francisco restaurants.[41]

In recent years, the quest for health has extended beyond the shark's fin (which is composed largely of cartilage) to its entire cartilaginous skeleton. Evidence is vanishingly thin that shark cartilage provides health benefits,[42] and those potentially positive effects that have been suggested appear to be matched by cartilage that's obtainable as a byproduct from meatpacking houses. But sharks, because their whole skeleton is cartilage instead of bone, give a much bigger yield of the product than do cattle or other mammals, from whose joints small amounts of cartilage must be cut and pried.

Today's shark-cartilage feeding frenzy goes back to 1992 and a book by William Lane and Linda Comac entitled *Sharks Don't Get Cancer: How Shark Cartilage Could Save Your Life*.[43] Actually, sharks have been shown to get all kinds of cancers, but at lower rates than do bony fish or mammals.[44] At any rate, the twin myths that sharks are immune to cancer and that we humans can share their protection by swallowing their cartilage have helped create a bull market in the stuff. At the top of that market in the late 1990s could be found a company owned by none other than William Lane. The FDA eventually pursued Lane Labs-USA, Inc. under the agency's "Operation Cure-All," charging that the company made spurious health claims for shark cartilage. In a settlement with the government in 2000, Lane agreed to stop making unsubstantiated claims and to pay a $1 million judgment, and in 2004, a federal court ordered Lane to stop selling shark cartilage and pay restitution to bilked customers.[45] Despite all that, it's still legal to sell shark cartilage. And there's still a lot being harvested and sold, either in general-health concoctions or as a source of chondroitin sulfate, which, in combination with glucosamine, is a popular non-prescription treatment for certain types of arthritis.

In a world increasingly permeated with the toxic fruits of "better living through chemistry" (see Chapter 9) we all could benefit from a bit more cancer insurance, whether it comes from green tea gardens or the deep blue sea. But if sharks were able to know what's going on, and if they could talk, they might ask, "What if all humans started consuming our cartilage?" Because commerce always runs ahead of science, they might not live long enough to learn the answer.

Hoodia. Three species of the plant genus *Hoodia* growing in southern Africa's Kalahari Desert contain a compound called P57 that's reputed to be an effective appetite suppressant. P57 was patented in 1996, and since the turn of the century, products made with it have been selling like low-carb hotcakes in the US.[46] Unfortunately, hoodia is a rare, apparently quite delicate plant growing in one of the world's most stressful environments. Despite a lack of scientifically valid studies showing that hoodia is effective, high demand has pushed it into threatened status. According to a South African news report in late 2006, hoodia "is now under threat after being hammered so hard by people trying to make a quick buck that it may become extinct within two years."[47] Nevertheless, promoters are expressing hope that hoodia harvesting will bring much-needed economic benefits to the region's poor people, as if such happy trickle-down benefits are ever really seen when there's a profit-driven rush on a biological resource.

In December, 2006, hoodia diet products were selling at $40 per ounce in the US, and exports of hoodia cut from the wild were approaching 500 tons annually.[48] Smuggling from South Africa and Namibia was thought to be rife. Attempts had been launched to grow hoodia plantation-style, but it was proving difficult to cultivate because of pests and other stresses. The coordinator of one such cultivation project told the *Los Angeles Times*, "It's an irony. It could be a way for people who feel they are overweight to help people who face a daily struggle to put something in their stomachs."[49] Or it might just be a way for the diet industry to make a few more fast bucks at the expense of a native ecosystem. Clearly, the number of the world's overweight people who'd be able to consume this threatened plant sustainably is nowhere close to "everybody."

Bottled water. The crystal-clear absurdities of bottled water have made it an easy target for writers and activists.[50] The contradictions are well known: an eco-conscious bottled-water industry achieves huge sales because industry and agriculture have polluted much of the tap-water supply; a product of heavy industry is promoted with images of bright glaciers and sparkling mountain brooks; a product consumed in order to sustain health reaches the buyer with the aid of the toxic plastic manufacturing and trucking industries; and much of this seemingly wholesome substance is supplied by transnational corporations that are looting water supplies from some of the world's poorest and thirstiest people, sometimes even trying to sell it back to them in bottles.

An estimated 1 billion people in the world have no access to potable water, and waterborne diseases kill 2 to 2.5 million people a year.[51] In 2000, the UN estimated that ensuring safe water for the billion people who need it would cost only about

$25 billion per year for eight to ten years.[52] That would seem to be an excellent investment; when it comes to drinking water, the question "What if everybody did it?" should not be a hypothetical one. But the global market's attention is elsewhere. Annual worldwide sales of bottled water are estimated at $50 to $100 billion—$11 billion in the US alone.[53]

A SELF-FATTENING INDUSTRY

This list of products has to be cut short at some point; otherwise, it could fill several books. As we leave this topic, however, it's important to emphasize that the dubious quality or performance of many nutritional commodities is of only secondary importance. Quacks and flim-flam artists have a long tradition of getting into the dieting and supplement markets, but neither better regulation of shady characters nor better research showing that their products really work will halt decimation of the planet's animal and plant species in the human quest for health. Plenty of products, including antioxidants and omega-3 fatty acids, have been shown to deliver solid results in at least some situations. And a lot of people really did lose a lot of weight on the Atkins diet. Whatever the commodities on sale, the results advertised—good health and long life—are always in high demand. Advertising neither sticks consistently to established facts nor conveys only pure fantasies; it's always a blend of the two. In the nutrition business, cruel scammers can be differentiated from admirable entrepreneurs only when they cross a certain fuzzy line.

Nevertheless, the fundamental principles of good health in human populations were worked out "on the ground," not in food factories or high-tech medical labs. Society should always have the right to develop approaches to good health that are rooted in the study of natural biological systems and don't necessarily have the seal of approval of the medical establishment. Many well-intentioned and knowledgeable people are doing good work in that area. Unfortunately, any of their work that shows profit potential ends up as fodder for the dieting/fitness/nutritional-supplement industry, whose necessarily all-consuming goal is to fatten *itself*. The bodies of its customers and employees figure in its calculations only as means to that end.

Agribusiness leaders often boast that Americans enjoy the world's cheapest and most abundant food. As millions of dieters can attest, that abundant food has turned out to be a badly mixed blessing for those of us who eat it. But as we'll see next, the consequences have been almost universally disastrous for food-producing landscapes, water bodies, people, and other animals.

5

"AGROTERRORISTS" CAN TAKE A VACATION

So vital is the dependence of terrestrial life on the energy received from the sun that the cyclic rhythm in which this energy reaches each region on the earth has gradually built itself through natural selection into the reproductive pattern of almost every species, vegetal or animal ... Yet the general tenor among economists has been to deny any substantial difference between the structures of agricultural and industrial productive activities.

Nicholas Georgescu-Roegen, *The Entropy Law and the Economic Process*[1]

For the life of me, I cannot understand why the terrorists have not attacked our food supply, because it is so easy to do.

Former US Secretary of Health and Human Services, Tommy Thompson, in his resignation speech, 3 December 2004[2]

Industrial or commercial output can be increased by building more capacity, stepping up the consumption of inputs, taking on more workers, and pushing workers harder and for longer hours. Farming, by contrast, is inevitably bound by the calendar—by month-to-month variation in the capacity of soil and sunlight to produce new plant tissue. It depends fundamentally on the productivity and the habits of non-human biological organisms over which humans can exert control only up to a point. Land, machinery, and workers all are forced into idleness for extended periods punctuated by times of frenzied activity. That clearly isn't the ideal pattern for efficient wealth generation, so the past century has seen relentless efforts to mold agriculture into the factory model as closely as possible and, where that can't be done, to graft more easily regimented industries onto an agricultural rootstock.

THE INDUSTRIALIZED FARM ECONOMY

Two small sets of numbers can illustrate the degree to which those efforts have transformed American agriculture. Table 5.1, using figures from the government's Bureau of Economic Analysis, shows that agriculture's contribution to the GDP, measured in real, inflation-adjusted dollars, has grown significantly over the past quarter-century but that the real action has been in industries that depend on the output from agriculture. The processing of raw agricultural products into food—what the government classifies as food "manufacturing"—has grown twice as fast as agriculture, and food service has grown almost four times as fast, much of that growth attributable to Americans' increasing tendency to eat out. Food processing and service are much more adaptable to organization along industrial lines, and the growth capacity that comes with that. Meanwhile, the value of food marketing—including packaging, labor, transportation, energy, advertising, profit, and other items—has risen even faster, reaching *four times* the value of the food produced on farms.[3]

Table 5.1 Contributions of three agriculture-related sectors to the United States' gross domestic product, in billions of dollars (adjusted for inflation), averaged over three 5-year periods.[4]

Period	Agriculture, forest, fisheries	Food manufacturing	Food service
1981–85	72	75	62
1991–95	96	122	116
2001–5	115	168	202

And the growth that has been achieved by the economy's agricultural sector has also gone, for the most part, to support industries whose output isn't physically bound to sunlight and the seasons. In Table 5.2, we see that the share of total farm output that goes to pay for industrial inputs—machinery, fertilizers, chemicals, etc.—has shot up from 30 percent of the average farm's total production in the 1940s to 50 percent today, while the share going to the farmers who raise the actual crops and animals has shrunk to less than half of what it once was. The dip in net income to 18 percent in the 1980s was a result of that decade's farm crisis. But production and consumption of purchased inputs continued to grow through the 1980s and beyond.

Table 5.2 Purchased farm inputs and net farm income expressed as percentages of US gross farm income, averaged by decade.[5]

Decade	Purchased inputs (percent of gross income)	Farmers' net income (percent of gross income)
1940s	30	47
1950s	38	35
1960s	43	26
1970s	45	23
1980s	47	18
1990s	49	22
2000s	50	22

Agriculture's built-in resistance to factory-style regimentation has not prevented a wholesale shift toward fewer and larger farms, to the point that 75 percent of economic output now comes from fewer than 7 percent of farms;[6] furthermore, there has been a steep rise in the proportion of farms owned by investors living in distant cities. And the necessarily Herculean efforts expended in attempts to break farming out of its natural cycles have been especially damaging. The well-used term "factory farming" may represent more an aspiration than an accomplished fact, but attempts to achieve it have been costly indeed.

Certainly, ecological devastation caused by cultivation and overgrazing is as old as agriculture itself. The list of civilizations that fell not to the sword but to the plow is long.[7] When Karl Marx and other thinkers of the eighteenth and nineteenth centuries discussed environmental crises, they usually emphasized the wearing-out of agricultural soils.[8] Soil degradation is a 10,000-year-old problem, inevitable as long as we continue supplying most of the human diet from annual grain crops.[9] But the "factory farming" that began in the twentieth century has accelerated the destruction. Attempts to concentrate and industrialize American agriculture have also cost it much of the resiliency that it had when it was a more highly dispersed activity. As a result, food production has become far more vulnerable to disruption by external forces, giving rise to a new fear-mongering term, "agroterrorism." Relying entirely on speculation about ways that politically motivated saboteurs might exploit modern agriculture's vulnerabilities, federal authorities are exhorting farmers to keep a round-the-clock lookout for agroterrorists lurking around fields or feedlots. Terrorists generally don't follow the tactical suggestions that government officials are constantly providing, but the obvious vulnerability of industrial agriculture does make official warnings sound believable.

Those who are sounding the agroterrorism alarm acknowledge that the increasing concentration of US agriculture, and its increasingly industrial infrastructure, are what make it more vulnerable. The US Government's General Accountability Office acknowledged in a 2005 report that

> the highly concentrated breeding and rearing practices of our livestock industry make it a vulnerable target for terrorists because diseases could spread rapidly and be very difficult to contain. For example, between 80 and 90 percent of grain-fed beef cattle production is concentrated in less than 5 percent of the nation's feedlots. Therefore, the deliberate introduction of a highly contagious animal disease in a single feedlot could have serious economic consequences.[10]

But the same, homegrown economic factors that have made agriculture and allied industries such a juicy target are already having consequences that are not easy to distinguish from the results of a still-hypothetical agroterror attack. Take a few common agroterror scenarios: terrorists might contaminate the food supply with biological agents; they could poison rural water supplies or release toxic clouds; they might breed bacteria resistant to most or all antibiotics; they'll release genetically engineered organisms; or—the all-purpose threat—they want to take away your freedom! As we will see, agrocapitalism has beaten agroterrorism to the punch on every one of those threats, and more.

HAZARDS OF FOOD PRODUCTION

Food poisoning. The CDC estimates that "76 million Americans get sick, more than 300,000 are hospitalized, and 5,000 people die from foodborne illnesses each year."[11] The flow of food contaminated with dangerous microorganisms in an ever-more-industrialized countryside is heavy and constant. Of the ten pathogenic organisms listed by the US Public Health Service as the most serious threats to human health, nine can be carried by meat and dairy products.[12]

Because the economy of animal-rearing is not quite as tightly circumscribed by natural cycles as is crop farming, factory-style methods have come to dominate meat, dairy, and egg production. Meanwhile, fish-farming proliferates. Concentrating and streamlining the means of producing animal products has made fortunes for a few, with predictable consequences for livestock and humans. The public-health threat that industrial meat processing poses to workers and consumers is well-documented.[13] Even the animals' diets are increasing the odds that consumers will fall prey to infection. Meat is frequently contaminated with feces as it leaves the slaughterhouse; cattle consuming a grain-based diet in feedlots (and that's the vast majority of beef cattle in America and Europe) are more likely to have the deadly bacterium *E. coli* 0157:H7 in their feces than are grass- or hay-fed cattle.[14]

Water and air pollution. A 2004 report by the Kansas Department of Health and Environment surveyed 19,500 miles of rivers and streams in the largely rural state. More than half of those miles—10,800—were "impaired for one or more uses" by pollution. Of more than 180,000 acres of lakes, 75 percent were similarly polluted. More than 40 percent of stream mileage and lake acreage was unable to "fully support" aquatic life, and 69 percent of lake acreage could not fully support domestic water uses. Agriculture was by far the biggest cause of damage to surface waters—exceeding industry, municipal discharge, sewage, urban runoff, mining, and oil drilling combined.[15]

Whereas gravity concentrates pollutants from croplands into rivers, lakes, and water tables, it's largely economic forces that have concentrated the pollutants from meat production. Feedlots, swine confinement buildings, and chicken houses are among the "hottest" sources of excess nutrients, especially nitrogen and phosphorus. Those nutrients, critically important to plant growth, move from the soil into crop, pasture, or range plants and from there into the bodies of animals. When the animals are kept for most of their lives packed into feeding facilities far from the landscape that feeds them, the nitrogen, phosphorus, and other elements in their urine and feces are transformed from important nutrients to troublesome wastes that often end up as water and air pollutants.

The USDA does a periodic census of the US farm animal population. In 2002, there were approximately 97 million cattle, 61 million hogs, and 9.6 *billion* chickens and turkeys.[16] Because of their huge numbers, and because farm animals produce up to three times as much wet-weight of urine and feces per pound of body weight as do humans, the total amount of waste produced is staggering: 900 million tons.[17] Between 1982 and 1997, the number of extra-large "farms" increased by more than half while the number of small farms fell,[18] and animal-raising has been geographically separated from crop production. Were today's animal populations scattered far and wide on crop-and-livestock farms across the nation, as they once were, farmland could re-absorb the organic matter and nutrients from the manure, to its benefit. But when large quantities of manure are produced in remote, densely populated animal-cities far from much of the cropped acreage, the soggy, bulky, heavy material cannot be hauled long distances economically. Therefore, 60 percent of the nitrogen and 65 percent of the phosphorus carried in the nation's manure production—that's hundreds of millions of pounds—is in excess of what can be absorbed by the farms where it's produced. Almost 20 percent of all the phosphorus—189 million pounds—is churned out by facilities in places where there's not enough cropland in the *entire county* to absorb it, even if every available acre were fertilized.[19]

When such gargantuan quantities of wastes are collected in one place, the potential for bacteria, protozoa, nutrients, hormones, and veterinary drugs, as well as arsenic, copper, selenium, and zinc, to escape from them into water supplies is very high. And that's exactly what happens, from household wells in Iowa to the waters of the Chesapeake Bay.[20]

Although more research is needed on the impact of polluted air from animal confinement facilities,[21] much recent work suggests that the odors coming out of such places not only assault the human nose but also indicate more serious health hazards. Children attending schools near animal-confinement facilities are at higher risk for asthma.[22] Hydrogen sulfide gas, with its characteristic "rotten egg" smell, is a common pollutant coming out of such operations, and even very small concentrations in the air can increase the rate of respiratory illness and even act as a neurotoxin.[23] Research has found that exposure to foul odors from swine facilities can lead to headaches, eye irritation, nausea, and even immune-system malfunction.[24]

Antibiotic resistance. Animals living in crowded, unsanitary conditions require frequent treatment for disease. In the US, antibiotics are injected into or fed to livestock at the first sign of illness, or even when the animals aren't sick at all, because the drugs promote weight gain and profits in healthy animals. Seventy percent of the antibiotics administered each year in the US are given not to humans but to farm animals, and one-fourth to three-fourths of those come right out again in urine and feces, polluting soil and water.[25] So-called "nontherapeutic" use of antibiotics simply to boost animal growth has been banned in Europe because it accelerates the development of hard-to-kill bacterial strains, but it's still allowed in the US. Studies have repeatedly detected bacteria with resistance to multiple antibiotics in the air and water around swine-feeding operations.[26] Resistant bacteria can escape from feeding operations not only through the air and water but also in end-products. Although tracking methods are never 100 percent certain,[27] an outbreak of antibiotic-resistant urinary tract infections of women in California in 2004 may have been caused by meat-borne bacteria from antibiotic-treated animals.[28]

Toxic chemicals. As writer Alexander Cockburn has pointed out, the post-9/11 grounding of all crop-dusting aircraft for a few weeks[29] was probably one of the few government actions during that period that actually protected citizens' health and lives. Of the 1.2 billion pounds of pesticides (fungicides, insecticides, herbicides, and others) consumed each year in the United States, more than 75 percent are used in agriculture.[30]

The consequences? By far the most comprehensive epidemiological study of the effects of agricultural chemicals is the National Institutes of Health/EPA Agricultural Health Study, which has been running since 1993. Scientists have been monitoring the health of private and commercial pesticide applicators and spouses—almost 90,000 of them so far. The still-unfinished research is suggesting that some agricultural chemicals present risks to humans. Preliminary results published in 2005 showed that, compared with similar populations of people not involved in pesticide application, the group studied had higher incidence of prostate, lip, gallbladder, ovary, and thyroid cancers.[31] As the massive study continues, the picture will become clearer. According to the project leader, "outcomes of concern" from pesticide exposure include cancer, neurologic diseases, reproductive problems, and respiratory diseases.[32] Recent independent research has shown a link between pesticide exposure and Parkinson's disease.[33]

Migrant farm laborers spend far more time in pesticide-drenched fields than do most other people in agriculture. In a single year, 1999, more than 60 million pounds of pesticides known to cause cancer or reproductive problems were applied in California, host to one-third of all hired farm workers in the US. Agricultural workers make up almost half of all pesticide-poisoning cases in the state, and that is probably an underestimate.[34] The majority of victims were poisoned not while applying pesticides but while simply working in fields. A 2003 review of migrant workers' overall health noted, with regard to pesticides, "Mass poisoning of hired farm workers continues to occur."[35] Chronic effects are not as easily traced as acute poisonings, but it is known that many Americans other than farm workers carry pesticides in their bodies. The CDC reported that in 2001–2, analyses of blood and urine samples taken from thousands of Americans between 6 and 60 years old commonly contained 22 different pesticides and pesticide metabolites.[36]

Genetic manipulation. More than 120 million acres of US farmland are sown each year to genetically engineered (GE) crops; that's more than half of the world's GE acreage. Since the mid 1990s, agribusiness has achieved an extraordinarily rapid adoption of such crops despite the vehement opposition of environmental activists. In their public campaigns, GE opponents have often put heavy or even exclusive emphasis on what is probably among the least of the threats it poses: the potential risk to human health.

Leading with the specter of "Frankenfoods" was a strategy based on sound public-relations logic. Raising alarms about the safety of food is always a surefire attention-getter and opinion-mover. Upton Sinclair learned that lesson when his great 1906 novel *The Jungle*, intended as a cry for justice on behalf of meat workers, instead prompted the passage of meat-inspection laws. Readers' revulsion over lack of sanitation in the plants that supplied their meat was far more powerful than their outrage over the desperate plight of plant employees. As Sinclair put it, "I aimed at the public's heart, and by accident I hit it in the stomach."[37]

If traditional food species like corn or soybean are engineered to produce plastics or pharmaceuticals and are eventually deployed over large acreages, the danger to our stomachs will then become clear. But as matters now stand, all the worry over hypothetical human-health impacts of currently deployed genes has distracted attention from the much clearer risks posed by pesticides, nitrates, biological pollution from animal-confinement facilities, and other of agribusiness's nasty products. That's not to say GE is safe; it most assuredly is not, but the greatest peril is economic, not gastronomic.

A 1980 US Supreme Court decision involving genetically engineered bacteria opened the door to the patenting of organisms, and in 2001, the Court upheld the validity of patents on crop varieties and all of their parts, including pollen, egg cells and genes. The effect of the Court's opinion, written by Justice Clarence Thomas, was to permit the draining of the agricultural gene pool into privately held ponds. Genes that existed many millions of years before the first farmer poked the first seed into the soil—and retain the ability to replicate themselves with no human assistance—are now bought and sold like any other commodities. Genetic engineering has been the biggest factor spurring the proliferation of patented genes and crop varieties, and that opens up fertile new territories for the high-input industrialization of farming.

Worst of all, patented genes help industry tighten its grip on what's left of the family farm. For example, the Monsanto Company's "Technology/Stewardship Agreement" is a contract that all farmers must sign when purchasing seed of soybeans, corn, cotton, sugarbeet, canola, or alfalfa. It dictates that when growing its resistant "Roundup Ready" varieties, farmers can spray only Roundup (the original brand of the herbicide glyphosate, developed by Monsanto and now off-patent) or a comparable herbicide authorized by Monsanto. It allows Monsanto to inspect and copy farmers' records and receipts; forms filed by the farmer with the US Department of Agriculture; and aerial photographs of the farm. And it prohibits the saving of seed harvested from a crop sown with Monsanto seed, meaning that farmers must return to purchase seed each time they plant.[38]

Monsanto reportedly has a division of 75 employees with a budget of $10 million devoted to investigating farmers' activities, and thousands of North American farmers are believed to have been investigated by the company or its hired private investigators, who have been accused of sneaking into fields to take seed or plant samples. By 2005, Monsanto was thought to have sued 147 farmers.[39] The company has a toll-free number (1-800-ROUNDUP) available to farmers for reporting suspected patent infringement by other farmers (a recorded voice advises: "For seed piracy or noncompliance issues, please press '2'").[40] Tipsters may remain anonymous if they wish.[41]

Industrial giants like Monsanto aren't the only players in the gene-patenting game. Since the 2001 Supreme Court decision, family-owned seed businesses have been jumping in, based on often absurd logic. One case involves a naturally occurring gene that can prevent the accidental movement of patented genes into organic crops, which, under federal standards, must not contain engineered genes. Wandering biotech genes have become a terrible headache for organic farmers. The natural tendency of pollen and seed to stray from field to field, along with improved genetic-detection methods, have made it harder than ever to produce certified organic food. So in 2002, a group of corn breeders led by scientists at Cornell University began developing organic varieties with built-in protection against stray genes. They took advantage of a naturally occurring gene that inhibits fertilization of a corn plant by uninvited pollen. It looked like a neat way to keep patented pollen out of the organic corn gene pool, but there was one hitch. The natural gene they wanted to use to keep out patented genes had itself been patented![42]

In April 2005, Nebraska seedcorn company Hoegemeyer Hybrids was awarded a patent describing the use of the gene to block foreign corn pollen. Novelty and "non-obviousness" are supposed to be two essential characteristics of a patentable idea. But an article published 50 years ago in *Agronomy Journal*, then the premier American journal of agricultural science, described the use of the very same gene for virtually the same purposes that are described in the Hoegemeyer patent.[43] The patent should never have been awarded, but until someone invests considerable time and money to mount a challenge, it will stay on the books. The patented system, dubbed Puramaize by Hoegemeyer, was scheduled for sale to farmers by 2008. Since 2004, the company has been conducting research with and providing breeding material to Swiss biotech giant Syngenta AG, via CHS Research, LLC.[44] The gene for protecting organic corn from GE crops could thereby end up one day being marketed by Syngenta, a global purveyor of GE crops.

WHO WANTS TO TAKE AWAY OUR FREEDOM?

The exploitation of people has always moved in parallel with exploitation of the land. In the words of Karl Marx:

> Moreover, all progress in capitalistic agriculture is a progress in the art, not only of robbing the labourer, but of robbing the soil; all progress in increasing the fertility of the soil for a given time, is a progress towards ruining the lasting sources of that fertility. The more a country starts its development on the foundation of modern industry, like the United States, for example, the more rapid is this process of destruction. Capitalist production, therefore, develops technology, and the combining together of various processes into a social whole, only by sapping the original sources of all wealth—the soil and the labourer.[45]

When the end-products of an industry are foods—essential products about which people always have strong feelings—it's too easy to look for both problems and solutions in the act of consumption and decisions of the consumer. But critics of capitalism, starting with Marx, have stressed that the crucial exploitation occurs not in the realm of the highly visible market (in the context of this chapter, the supermarket) but out of sight of all but a few, in the workplace. That is where labor power and the Earth's resources join forces to produce the extra value that supports the capitalist enterprise. In that workplace, it is the owner, not the worker, who makes the decisions. Although, or maybe because, it is so difficult to mold agriculture into a factory-like system, it is in agriculture that some of the cruelest exploitation of workers occurs. Because animal and meat processing can be better squeezed into the classic industrial model than can many other activities that used to happen on family farms, exploitation in that industry is routine. Relentless intensification and refinement of processes have greatly increased productivity, but at the cost of confining humans in conditions almost as bad as those faced by the animals they process.

In 1970, the number of broiler chickens living in the United States was 15 times the number of people. By 2005, when the human population had grown to almost 300 million, there were 30 times as many broilers.[46] It was Americans' hunger for healthful, low-fat, and in later years, low-carbohydrate meals that spurred poultry production. In a 2004 survey of Americans who said they were "influenced" by low-carb diets, 67 percent said chicken was the "most appropriate" for a low-carb, low-fat diet, compared with 14 percent for beef and 9 percent for pork. Said a spokesperson for the company doing the survey, "Chicken's long-term reputation for lower fat content is clearly still an asset."[47] But, as Table 5.3 shows, the astonishing increase in poultry consumption has led to no discernible improvement in Americans' health.

Table 5.3   Poultry consumption,[48] waist measurement,[49] and blood-pressure treatment[50] for the average American. Comparable waistline statistics are not available from the 1970s.

Decade	Per-person poultry consumption, lbs per year	Average waistline, age 20 or greater, in inches	Percentage of adults taking blood-pressure medicine
1970s	52	–	10
1980s–90s	79	36.6	11
2000s	101	38.8	16

Meanwhile, the workers who turn live birds into breasts, thighs, and nuggets have seen their own health go downhill. We all talk about "the environment," but there really is no "the" environment. Day to day, the specific environments that people experience span a huge range. And there's no way to kill and dismember animals in pleasant, comfortable surroundings.

The transformation from farming to manufacturing has gone farthest of all in poultry production. Farmers have become no more than the caretakers of birds, feed, and other inputs that are all owned by the corporation with which they contract, while "chicken-catchers" on the farm and workers in the processing plants have been converted into components of a vast breast-and-wing manufacturing machine. Because the "units" being handled—the birds—are so much smaller and more numerous than cattle or hogs, a steady, ample supply of human labor is crucial to the poultry business. That puts even more pressure on its owners and managers to hold down the cost of each human being employed.

One of the most commonly reported health problems among poultry workers is repetitive-motion injury. In 2002, *Fortune* magazine reported that while the overall injury rate in poultry plants was more than double the average for private industries, poultry workers are 14 times more likely to suffer debilitating injuries stemming from repetitive trauma.[51] Look through the 2005 Human Rights Watch report *Blood, Sweat, and Fear: Workers' Rights in U.S. Meat and Poultry Plants*[52] (which demonstrated that everyday conditions in plants from North Carolina to Nebraska to Arkansas violate a host of international human-rights standards), or Eric Schlosser's 2001 bestseller *Fast Food Nation*,[53] and read the ghastly stories of workers' lives in the industries that process poultry and other meats. Then pick one of the jobs they describe, a single task, and try to imagine repeating it all day long, as quickly as you can.

Better yet, practice to be a poultry worker yourself. Take two 5-pound weights, hang them on hooks above your head, pull them down, and then repeat the cycle maybe 15,000 times in an 8- to 12-hour day. Be sure to do this in the dampest possible conditions, with the temperature either above 90 or below 50 degrees Fahrenheit. You'll just have to imagine that the weights are panicky live animals with beaks and claws. If you need a break, you might try cutting half-frozen chickens apart with dull scissors for a while. If your hands, wrists or back don't ache the next morning, repeat the process five or six days per week until they do. Don't worry—they will.

Repetitive-motion injuries are common in the slaughtering and processing industries, but they are epidemic in poultry work. Very large numbers of the relatively small animals pass through a plant each hour, requiring workers to repeat their actions more often than in beef or pork plants. On a recent visit to my hometown of Gainesville, Georgia, which since World War II has been at the center of one of the nation's top poultry-producing regions (and which sports two downtown monuments, one of a Confederate soldier and the other topped by a statue of a broiler hen), I asked Dr. Nabil Muhanna, a local neurosurgeon, about a condition he treats that is ubiquitous in poultry work: carpal-tunnel syndrome. He pointed to his own wrist to show me where the median nerve passes between the bones of the wrist on one side and the transverse carpal ligament on the other. That is, he said, how the nerve is naturally "packaged."

But the hand and arm motions required of poultry and meat workers, if they're repeated enough times, cause inflammation of the nerve. The resulting severe pain can be relieved by cutting through the ligament to loosen the packaging, but the wrist remains weak. For decades, Dr. Muhanna has been slitting the transverse carpal ligaments of poultry workers, as well as performing surgery to relieve the misery associated with other of their work-related conditions, such as lumbar stenosis and disk disease.

I asked Dr. Muhanna to help me interpret a figure I had seen—a claim that 15 percent of workers in the poultry industry suffer repetitive-motion injuries. When he reacted with sharp disbelief, I thought he was going to say the figure is exaggerated. Instead, he said, "Look, if you work at hanging and cutting chickens long enough, you will inevitably get carpal-tunnel."[54] And other repetitive motions, like squatting, will always create health problems if done constantly, he said. If Dr. Muhanna is right, many workers either have unreported—and probably untreated—injuries. Many simply haven't stayed on the job long enough to succumb; the industry has a notoriously high turnover rate.

A 2007 study compared the prevalence of musculoskeletal symptoms in almost 300 poultry workers in northeastern North Carolina (all women, 98 percent of them African-American) with the prevalence in a similar group of women in the same area who had other occupations. Between two and three times as many of the women working in poultry plants suffered neck, shoulder, and back pains as did workers in other jobs. Fully four times as many poultry workers—36 percent of them—had hand and wrist problems. The authors of the study noted that high injury rates arise partly because of "growing preferences for prepackaged parts and deboned meat" leading to emphasis on "the rapid slaughter, disassembly and packaging of birds, with the work characterized by rapid line speed and extreme division of labor."[55]

Steve Striffler, an associate professor of anthropology at the University of Arkansas, has seen the pervasiveness of carpal-tunnel firsthand. He worked "undercover" for two summers in poultry-processing plants and wrote a book on industrial chicken.[56] Striffler told me that "virtually all the women who worked on my line wore some kind of device because of problems with their wrists—all due to the repetitive nature of the job."[57] In an article recounting his experience on the so-called "saw line" in a Tyson poultry plant, Striffler described what happened when the plant upgraded the electricity supply to the breading machine that he operated, solving the problem of repeated breakdowns, which had plagued the saw line up to that point:

> For the rest of the workers, particularly those on-line, it is devastating. The lines shut down less, there are fewer breaks, and the pace is quicker. By the end of the week, Blanca, a Mexican woman in her 50s, is simply overwhelmed. She has been hanging chickens for too many years, and her body simply cannot stand the faster pace of the production line. Hoping to stay at Tyson till she retires, Blanca is forced to quit within the week.[58]

Since the 1980s, ownership in the meat and poultry industries has become heavily concentrated in the hands of a few corporations, and competition among those behemoths is fierce. To cut costs, companies are squeezing maximum output from a minimum number of workers. That means more wear and tear on human bodies, with predictably grim consequences. A vivid illustration of the link between company greed and workers' injuries can be found in an appendix to the Human Rights Watch report *Blood, Sweat, and Fear*, in which are reproduced portions of a 2003 federal court transcript in which a US attorney questioned a Tyson manager:

Q: After you got to [the] Glen Allen [, Virginia, poultry plant], were there any major processing changes?

A: Yes. We went to the 50 degrees, which eliminated washdown time.

Q: When you say you went to 50 degrees, I think you're going to have to explain that a little more.

A: We lowered—what we did was we lowered the temperature of the plant in order to eliminate a midshift washdown because by lowering the plant temperature, what you do is slow bacterial growth, therefore, it eliminates a need for washdown, and, therefore, more production can be achieved.

Q: What had the temperature in the plant been before that change?

A: Sixty plus degrees. High sixties.

Q: Did Tyson just turn the thermostat down?

A: No, sir. It was a major expansion. And they had to install the HVAC machines, like massive air conditioners ... to make it the desired temperature to meet USDA specifications to avoid a washdown. ...

Q: Did that change in plant temperature and working conditions have an adverse effect on staffing at the Glen Allen plant?

A: Yes, sir.

Q: How?

A: It sure did, because what it was doing, it was taking our senior workers, who had been there for quite a number of years, and it was making—it's so hard on them, they were complaining of bursitis, arthritis and increased musculoskeletal problems...

Q: What about workers' compensation claims?

A: We didn't have to worry about workers' compensation claims with any of the USA Staffing personnel [i.e., workers hired through an outside contractor].[59]

Workers' compensation laws make companies accountable for damage to the health of their employees when it happens at work. Ever conscious of their bottom line, corporate officials will try to minimize what they pay out for, say, a severed limb. But if it's the less visible damage caused by repetitive motion, stress, and strain, they may well try to avoid responsibility altogether.

Maria Chavez suffered repetitive-motion injuries while working at a Tyson poultry plant in the early 2000s. Although she has undergone wrist-replacement surgery on her left wrist, it is still in constant pain. And now her right wrist frequently goes numb. Speaking through an interpreter, she told me that over time, her tasks have included making boxes, packing boxes with chicken and taking frozen chickens apart. I asked if she used knives or scissors to disassemble the chickens. "Neither," she said. "I used my bare hands." She was also assigned the job of clearing out a large weighing scale that, she says, clamped down and cut into her already-injured wrist. The company nurse had her soak the wrist in hot water for 15 minutes and then return to work. Tyson refused to comply with a court order to pay for her medicine, physical therapy, and surgery.

Maria's husband Manuel worked at the same Tyson plant from 1991 until 2006, doing mostly heavy mechanical work. He said that for two or three months after he first reported severe back pain, the only treatment he received from the company nurses were hot patches and Tylenol. He was finally permitted to have a local doctor treat the pain under Tyson's workers' compensation plan, but, he says, the company ordered him back to work too soon, telling the doctor he'd be put on light duty. In reality, he says, "I was put back on all the heavy jobs, whether I could do them or not." His doctor told the company flatly that he should not be doing such tasks. What was Tyson's response to its employee of 14 years? "They called me a drug addict," Manuel says, "saying I was faking the pain to get the painkillers."[60] The Chavezes filed a lawsuit against Tyson and eventually obtained a settlement. But, their attorney told me, they succeeded only because they were able to obtain video footage of Manuel doing the strenuous jobs that the doctor had insisted he not be forced to do.

In that other top poultry-raising region, northwest Arkansas, the economy has never been stronger. The headquarters of the world's biggest retailer, Wal-Mart Stores, Inc., is located in folksy Bentonville (its picturesque little town square festooned with dozens of American flags, with a statue of a Confederate soldier at the center) and the home office of top meat producer Tyson Foods is just down the road in Springdale. But a couple of hours farther south, in and around Danville, life is harder. Many of the residents are recent immigrants, and their biggest employers are poultry processing units run by several companies. In 2006–7, Miranda Cady Hallett spent a lot of time with poultry-plant employees who'd come to the Danville area from El Salvador. She was a student in anthropology at Cornell University, conducting research for her Ph.D. dissertation on immigrants in the poultry industry.

I asked Hallett which assignments in the plants are considered by workers to pose the greatest direct environmental hazards. One such job, she said, was the midnight-to-6 a.m. sanitation shift, in which workers pressure-wash everything in the plant with a strong solution of extremely hot water and chlorine bleach: "their protective hoods fog up almost immediately, which leads to slipping accidents, so they often don't wear the hoods. That's how they sometimes get their faces burned. People who work there say that those who take the sanitation shift stand out among the workers because they tend to have blotchy, mottled faces. I don't have medical evidence of this, but it's something that's accepted among the workers."[61] Studies have shown that prolonged exposure to chlorine bleach solutions is associated with skin damage and asthma.[62]

Hallett said single mothers often take the sanitation shift, risking their health in order to be with their children during the day. But, she told me, "the companies try to avoid responsibility for health care at all costs. They will fire people to avoid paying. They can afford to keep increasing lines speeds because the additional profits outweigh the costs of injuries, which they often don't pay for anyway. Workers are treated as disposable bodies." One worker told Hallett (asking her not to use his or his employer's name) that he had been forced to complete a shift after showing his supervisor his broken finger, which clearly veered off at an unnatural angle. Hallett said employees tend to avoid confrontations over such mistreatment, because of their immigrant status. But after the broken-finger incident, plus an earlier sprained ankle and company attempts to avoid full compensation, this worker had had enough. He told her, "I went in and gave my supervisor's supervisor and the company nurse a real talking-to. I said, 'I was in the Salvadoran Army, and even *they* didn't treat us this badly!'"

Dr. Muhanna, the Georgia neurosurgeon, has seen what eventually happens to people who continue working in the poultry industry despite their pain and their employer's intransigence: "They become washed out of their humanity. Lives and families are devastated. If the company can demoralize, harass or degrade an injured worker, they will do it."

But impoverished people from around the world don't have to move to Georgia or Arkansas to have their lives transformed for the worse. As the next two chapters will show, the long arm of Western industry can reach them right on their own farm.

6

HUNGER FOR NATURAL GAS

Visualize, if you will, a cut of beef roasting in a suburban kitchen somewhere in the United States. The oven's flame is produced by natural gas that's being piped into the kitchen by a utility company. If it's summer, the central air-conditioning system, powered by electricity from a natural gas-fired plant, is locked in thermodynamic battle with the oven. The meat in the oven, compliments of a feedlot-raised steer, is made largely of proteins containing nitrogen from synthetic fertilizers. The nitrogen in the fertilizer was pulled out of the atmosphere by an industrial process that uses natural gas. After dinner, an automatic dishwasher will clean the pan and other dishes using electricity from the gas-fired power plant and hot water from a gas-fired tank in the basement.

As fossil fuels go, natural gas provides relatively clean and efficient energy for cooking. But the other uses of gas in the scene above represent mostly luxury consumption. A kitchen doesn't have to be kept at November temperatures in July. With installation of a relatively simple apparatus, sunlight can efficiently heat water, and dishes can be hand-washed. And there would have been plenty of room on America's vast, fertile landscape for the steer that provided the meal to spend its life grazing on grasses and legumes, rather than milling around in a feedlot, eating corn raised on natural gas.

Now consider another image: a pot of rice cooking outside a hut in a village in southern India. The heat to boil the water is produced with wood or cow dung or kerosene. The rice will provide some protein to the family that eats it, but to get the right balance of proteins, they will also need peas, beans, or yogurt as well. The cooking pot will be washed with cold water and dried in the sun. But unlike the American steer from which came the roast, the rice paddy that provided the centerpiece of this meal could not have functioned without nitrogen fertilizer derived from natural gas. The quantity of natural gas required is tiny compared with that consumed by the roast-beef meal and the kitchen in which it was cooked, but without gas the rice in this meal might never have existed.

The Indian family that will eat the meal consumes a lot less natural gas than do American families, but what they do use, via nitrogen fertilizer, is fundamental to their very existence. That leaves them exposed to the vagaries of the global energy market. There, the need for basic food production can easily be out-bid by the desire of affluent societies for climate-controlled homes, arsenals of electric appliances, and 24-hour hot water.

We're all familiar with the standard environmental-disaster scenario: A devious corporation is knowingly and illegally releasing a toxic substance into the air and water, or perhaps ruining a forest or wetland; eventually, the perpetrators are exposed and brought to justice by courageous victims, lawyers, and/or investigative journalists. That usually makes for a gripping story, but it's not the story of nitrogen fertilizers and natural gas. Those substances manage at the same time to be both essential friends and toxic foes. Their relationship with humanity and the rest of the planet is complex and often ambiguous, but it's not the result of any dark conspiracy. Much more than the standard scenario, natural gas and nitrogen illustrate the tangled economic and ecological interactions that are deciding our fate under global capitalism.

NITROGEN, HUMAN EXISTENCE, AND ECONOMIC LOGIC

Crop plants assemble carbon, hydrogen, oxygen, and nitrogen into proteins that are essential both to plant growth and to the diets of humans and other animals. Of those four elements, nitrogen is the one that's too often in short supply. Yellowish, stunted crops, whether they're in an Indiana cornfield or an Asian rice paddy, can most likely blame their anemic complexion on a lack of nitrogen.

Before the rise of large-scale industry and agriculture, the biosphere got its nitrogen mostly from specialized microorganisms that are able to take pure nitrogen gas from the air and "fix" it into compounds that other living things can utilize. Most prominent among such microbes are bacteria of the genus *Rhizobium* that colonize the roots of plants like beans, peas, alfalfa, tamarind trees, and a multitude of other plants belonging to the legume family. But biological fixation and other natural processes are simply not capable of supplying nitrogen-rich protein to everyone on Earth. The prolific growth of the human population has been made possible only by the widespread application of industrially produced nitrogen fertilizer. Our species has become as physically dependent on fertilizer, and therefore on fossil fuels, as it is on soil, sunlight, and water.

Resource expert Vaclav Smil of the University of Manitoba has demonstrated in stark terms just how deeply addicted to synthetic nitrogen the global food system has become.[1] Much of the protein in human bodies, planet-wide, was built from synthetic nitrogen that has been applied to crops over the course of the past century. Using straightforward math, Smil argues that without the use of industrially produced nitrogen fertilizer, about 2.6 billion people out of today's world population of 6.6 billion would never have existed. Smil's related calculations show that if farming depended solely on naturally occurring and recycled nitrogen fertility, the planet's cropped acreage could feed only about 50 percent of the human population at today's nutrition levels, which are higher now than during those earlier decades of population growth.

The momentum of past population growth is expected to put two to four billion more people on the planet by 2050, even with concerted efforts to rein in reproduction. Almost all of that increase will occur in Africa, Asia, Latin America, and the Middle East. That will double the demand for nitrogen fertilizer in those regions, and by that time, says Smil, 60 percent of their inhabitants will depend existentially (in the literal sense, not the philosophical one) on nitrogen fertilizer made largely with natural gas.

But dependence on synthetic nitrogen is geographically lopsided, concentrated largely in impoverished nations that have large populations relative to their usable cropland. One analysis of the world situation concludes that "while some countries are fundamentally dependent on synthetic N for food production, many countries have the capacity to greatly reduce or eliminate dependence on synthetic N through adoption of less meat-intensive diets, and reduction of food waste."[2] For instance, the world's third most populous nation, which also happens to be the third biggest nitrogen fertilizer consumer, could conceivably do without the stuff altogether. If that country, the United States, were to reduce its meat consumption to levels recommended for good health; raise all livestock on pasture and rangeland instead of pouring fertilizer onto corn and other grains; rotate harvested crops of grain with non-harvested crops of nitrogen-fixing legumes grown only to improve soil fertility; put more land into efficient, deep-rooted perennial plants and less into inefficient annual crops; curb the waste of foodstuffs (sometimes estimated at 25 percent of total production); and cut grain exports, the US could maintain its food supply without using any synthetic nitrogen at all.[3]

That, unfortunately, is not the preferred way of growing food in our economy, because to do so would eliminate a whole series of profit-making opportunities. Synthetic nitrogen is part of a hot-and-heavy nutrient market that augments the profits of petroleum companies, fertilizer producers and applicators, seed dealers, farm implement manufacturers, grain traders, meat packers, millers, bakers, food wholesalers, supermarket chains, restaurants, advertising firms, hospitals, insurance companies... as we've seen, the list includes almost everyone but the farmer. It doesn't matter that this system automatically funnels a big share of valuable nitrogen compounds directly from the factory where they're produced to bodies of water where they're not wanted; it's the sheer throughput that generates the wealth.

GAS: SO GOOD IT'S BAD

In that vast volume between the Earth's surface and the atmosphere's upper limits, nitrogen is the most abundant element. We're continuously bathed in nitrogen gas, which makes up 78 percent of the air we breathe. But in the air, nitrogen atoms are paired up, each atom linked to one other by an extremely tight "triple bond." Those molecules can't be used by living organisms unless that bond is broken, and only a tiny percentage of microscopic species have developed means to do that.

Prying nitrogen atoms apart without the help of those miraculous microbes requires massive doses of energy; it happens, for example, around a bolt of lightning. Eventually, with energy from fossil fuels, *Homo sapiens* became the first non-microbial species to fix nitrogen. The fateful event occurred in 1909 when two German scientists, Fritz Haber and Carl Bosch, developed an industrial method to reassemble nitrogen atoms into another molecule, ammonia, on a mass scale. Ammonia and compounds made from it are usable by crop plants and form the foundation of synthetic fertilizers.

The three inputs to the Haber-Bosch process are nitrogen, hydrogen, and energy. Currently, fossil fuels are used to produce the hydrogen and energy. Therefore, to extend Vaclav Smil's reasoning, 40 percent (soon to be 60 percent) of the Earth's inhabitants owe not just their lifestyles but their very lives to the daily consumption of nonrenewable resources: natural gas, and, to a lesser and decreasing extent, oil and coal.

In 2004, Julian Darley published his book *High Noon for Natural Gas*, in which he argued that the era of cheap and plentiful gas, like that of cheap oil, is coming to a close. Humans began tapping the Earth's deposits of oil and natural gas a little over a century ago. We've been exhausting the planet's oil reserves more quickly than its gas reserves, because for many purposes oil is easier to transport and use. The planet's gas reserves will last longer than those of oil, but the world is now using more gas each year than is being discovered—an ominous sign.

Accelerating consumption across the globe, wrote Darley, will continue to drive up natural gas prices, deplete reserves, and trigger chronic shortages. In a world where growing energy demand has begun to run up against environmental limits, gas is almost too good to be true, and, it seems, too good to leave in the ground. The pressures squeezing more and more gas from the earth are diverse:[4]

- Countries trying to meet the greenhouse emissions limits set by the Kyoto Protocol are rapidly building natural gas-fired power plants, which emit much less carbon dioxide than do coal plants. In the United States, the world's number-one greenhouse deadbeat, we're now building mostly gas-fueled power plants, just as our domestic gas reserves dwindle.

- In response to international criticism of its heavy coal use, China intends to triple or quadruple its consumption of natural gas for power generation in the coming decade.

- The petroleum industry is pushing hard to build large numbers of liquefied natural gas (LNG) tankers, along with the requisite high-tech port facilities in the major producing and consuming nations. That will make it easier for a big energy-using nation like the US to suck not only from gas pipelines on its own continent but from wells almost anywhere on the planet, as we currently do to feed our oil habit.

- Building and operating a global LNG system will require vast amounts of energy—much of it supplied by gas, of course. To produce the power required to haul liquefied gas across oceans while keeping it cooled to about -260 degrees Fahrenheit, LNG tankers must draw on their own gas cargo. Such a system is far more fragile than the lower-tech energy systems we're used to; an explosion at a LNG terminal could produce a fireball a mile wide—qualifying LNG as a potential "WMD."

- The process of extracting heavy oil from sands in the Canadian province of Alberta—often looked to as a key new resource in a "safe" part of the world—requires natural gas, and a lot of it. Darley predicts that if the oil sands are to satisfy even one-eighth of North America's demand, they will have to absorb a quarter to a half of Canada's natural gas production!

- Hydrogen is often hailed as a fuel of the future, but today, most hydrogen is manufactured from—what else?—natural gas. As we'll see, alternative methods of extracting hydrogen are more complicated and costly, so natural gas will likely continue as the preferred source.

Not everyone is as pessimistic about natural gas as is Darley. The US Department of Energy, for example, paints a much rosier picture of potential gas reserves.[5] But on one point there seems to be universal agreement: Consumption of the world's natural gas will continue to accelerate, and in the rush, gas could prove even more volatile than oil, politically and economically as well as chemically. The timetable for a peaking or plateauing of natural gas production and its eventual decline is even harder to forecast than it is for oil. But a perfect storm of long-term forces appears to be blowing demand in only one direction—up—and access to such a resource will likely go to those players with the most money and the strongest armies.

Why armies? Because the world's remaining natural gas reserves lie mostly in the Mideast, Central Asia and Russia, almost guaranteeing that a century of conflict and chaos lies ahead in those already unsettled regions. Other reserves are scattered among a number of small countries, many of them in Africa. Tapped judiciously, their gas could sponsor decades of domestic fertilizer production for those nations. But, as people from Kirkuk to Caracas to the Niger Delta can tell you, fossil fuel reserves also can attract a lot of unwelcome attention from more powerful, energy-hungry nations.

COAL: A LOUSY PLAN B

It's possible to acquire the hydrogen and energy needed for manufacturing ammonia (and from it, nitrogen fertilizers) without tapping natural gas reserves. The most widely discussed alternative fuel is that old-time favorite, coal. Coal can serve as an energy source for producing a gas—hydrogen or a hydrogen compound—that, in turn, can be used to fix nitrogen into ammonia. Coal as a source of gaseous fuel has been around a long time; so-called "town gas" derived from coal lit up urban areas around the world in the nineteenth and early twentieth centuries. And China's nitrogen-fertilizer industry, at one time almost exclusively coal-based, continues to rely on coal for 60 percent of its ammonia production.[6]

In recent years, a wide variety of increasingly more sophisticated and cleaner coal-based methods have been proposed, and some have been employed on a small scale. The most advanced methods of coal gasification involve a complex series of chemical reactions.[7] Inputs are primarily carbon from the coal and oxygen from the atmosphere; the main end-products are hydrogen and carbon dioxide. The hydrogen can be used as a fuel itself, or to produce other fuels, or in the Haber-Bosch process to produce ammonia. If purified oxygen rather than plain old air is used to burn the coal, the carbon dioxide that comes out is in a highly concentrated form that can be captured and pumped underground, where it won't contribute to global warming.

That all sounds nice, until we remember that twenty-first-century society was built on a bonanza of fuels like oil and natural gas that can simply be pulled right out of the earth, with the wastes from refining and burning them simply dumped into the atmosphere. Had we instead been forced to rely on fuels synthesized from dirty, carbon-rich coal or tar sands or other much-touted fuels of the future, and had we mounted the herculean effort required to keep that carbon out of the atmosphere, we'd be living in a very different world today. Consider what will be necessary to replace natural gas with "clean coal" for production of fuel and nitrogen fertilizer: [8]

- *Multiplying the horrors of coal mining.* Bringing coal up from ever-deeper subterranean seams continues to be the worst sort of exploitation of human labor. Coal-mine accidents kill about 6,000 workers per year in China.[9] Strip mining, especially in the truly monstrous form of mountaintop removal, is an ecological catastrophe in the Appalachian mountains of the US. Some have touted the use of waste coal for gasification, but if it's to substitute, increasingly, for gas and oil in supporting growing economies, every scrap of waste coal will be scooped up in short order, and industry will turn quickly back to the mines. Finally, a lot of energy is needed to haul heavy coal from the mine to the ammonia plant.

- *Building and maintaining yet another vast, complex, costly, and potentially fragile infrastructure.* Getting natural gas, high in hydrogen and low in carbon, from a well-head via a pipeline is the kind of simple system that can really keep a growth economy humming. As we have seen, hauling it between continents in its liquid form, LNG, is more costly and dangerous but still simple compared with proposed systems to use coal in its new, "clean" incarnation. "Clean coal" technology takes a long, winding road to the finished product, one that's unproven on a large scale and sure to be costly in money and energy.

- *Stepping down to a less energy-efficient technology.* Making nitrogen fertilizer with coal rather than natural gas uses much more energy (and produces much more carbon dioxide) per ton of product. That is why only two of the 32 ammonia plants imported by China since 1973 use coal as the feedstock,[10] while 79 percent of India's ammonia capacity installed since 1980 uses natural gas, and only 5 percent uses coal.[11] One report estimated that in India of the early 1990s, coal-based production of ammonia was using more than four times as much energy per ton of product as was production using natural gas.[12]

- *Going to extraordinary lengths to keep "clean coal" clean.* In a greenhouse-conscious world, natural gas owes its great popularity to its low carbon content. But coal gasification requires strenuous efforts to keep control of the huge bulk of carbon that comes out of a coal mine. To generate the input of pure oxygen needed for carbon-capture may take 20 percent of the coal's energy right off the top. And no one has figured out how to store large amounts of carbon dioxide underground, safely and forever—certainly not the colossal amounts that full-scale gasification would produce. It is possible to get the carbon dioxide down there; currently it's pumped into some oil wells to help push the oil out. In other words, it's economical to hide carbon dioxide underground as long as you're using it to get a carbon-laden fossil fuel aboveground! The renewed enthusiasm for coal gasification is, more than anything else, evidence of the growth economy's growing desperation.

Nitrogen-rich ammonia could be produced in a solar future, by using big doses of wind- or photovoltaic-generated electricity to split hydrogen atoms away from water. When the wind is really howling but demand for electricity is down, a wind farm equipped with an ammonia plant could make good use of energy that would otherwise be wasted.[13] But, compared with pumping hydrogen-rich natural gas right from the ground, the huge expense and effort required to obtain hydrogen from water (using either new wind and solar technology or old coal) will result in intense competition among its many potential users for every molecule of the stuff. The advantages of "natural" natural gas over any of its competitors have guaranteed it an enthusiastic market right up until the day the last gas well hisses its last hiss.

FULL JACUZZIS, EMPTY STOMACHS

Fertilizer production currently uses only about 5 percent of the world's natural gas production, and nonagricultural uses are already asserting greater dominance over tightening gas supplies in the United States. The escalation of natural gas prices in recent years has made fertilizer production far less profitable; as a result, the US has lost 30 percent of its nitrogen fertilizer production capacity.[14] American farmers now obtain more than half of their nitrogen fertilizer from abroad, making them the world's biggest importers of the product.[15]

Meanwhile, as natural gas becomes both more portable (via LNG) and more essential to food production in much of the world, impoverished farmers in Indonesia and Egypt will find themselves bidding for it against wheat farmers in Washington State, homeowners in sweltering Phoenix or frigid Buffalo, and appliance exporters in Shanghai. In a cruel irony, gas is most crucial to feeding people in those very countries where poverty keeps overall per-person gas consumption low. While only 5 percent of world natural gas consumption goes to make ammonia,[16] the figure is 40 percent in India.[17] In contrast, the United States' nitrogen-fertilizer industry is based almost exclusively on natural gas, but fertilizer accounts for only 3 percent of total US natural gas consumption. US households consume 4,840 trillion Btus directly and another 780 trillion Btus in the form of gas-generated electricity. (Almost 20 percent of US electricity is produced with natural gas, and that proportion is increasing.) Table 6.1 shows what all that gas does for American homes.

Table 6.1   Total annual quantities of natural gas consumed by US households for various purposes, either directly or through electricity generated with natural gas[18]

Energy source	End-use	Natural gas consumed (trillion Btus per year)
Direct from pipeline	Home heating	3,340
	Water heating	1,162
	Other gas appliances	339
From gas-fired electricity	Air conditioning	125
	Refrigerators	107
	Other kitchen appliances	74
	Home heating	79
	Water heating	71
	Lighting	69
	Clothes drying	45
	Other appliances	180

US residential consumption alone exceeds the residential, commercial, and industrial gas consumption of China, India, Indonesia, and Pakistan combined. Those are the world's first, second, fourth, and sixth largest nations, home to almost half of humanity. Of course, in the US, space heating is the biggest residential use, and that can be as essential to preserving life as is food production. Enormous decreases in natural gas consumption could be achieved through better insulation and lower thermostat settings, but enormous demand would remain.

Dr. S.P. Wani is a Principal Scientist at the International Crops Research Institute for the Semi-Arid Tropics in Hyderabad, India. He studies, among other subjects, the quantities of nitrogen fertilizer needed by grain crops and the quantities that farmers can afford to apply. He has conducted research on small farms across India, from the state of Rajasthan in the dry northwest to Tamil Nadu in the humid southeast. What proportion of those fields has he found to be deficient in nitrogen? "All of them," he told me. "The typical field is getting maybe 25 percent as much nitrogen as it needs." Even irrigated rice paddies belonging to prosperous farmers along India's east coast, he said, get "only 60 percent or so of what they need. The factor limiting crop yield is always a deficiency of nitrogen or water or both."[19]

What will happen if natural gas prices shoot to high levels and stay there? Wani sees trouble: "Farmers are buying as much nitrogen as they can afford. At this stage, they are just at the break-even point. If fertilizer prices go up, unless either their yields or government subsidies also go up, they are in deep trouble." He confirmed that crop production in India is heavily dependent on natural-gas-derived nitrogen, because biological sources like legumes can provide only a fraction of what's needed, "and organic fertilizers are much more costly than synthetic ones. They're used on high-value crops like vegetables and fruits, and that drives up the price for everyone."

NITROGEN: TOO LITTLE, TOO MUCH

Across India and other nations of the global South, farmers walk through their small plots with a panful of urea or other synthetic fertilizer, metering out the precious granules by the handful, to stretch the supply as far as it can go. Meanwhile in America's

Corn Belt, hulking farm implements put luxurious excesses of nitrogen into the soil. Much of that nitrogen bypasses the crop and moves quickly into streams and groundwater. A 1998 CDC report showed that, as a result, 15 percent of domestic wells in Illinois, 21 percent in Iowa, and 24 percent in Kansas were contaminated with nitrates above a safe level, and the situation has not improved since.[20] Many, but not all, studies have shown links between nitrates and various cancers in adults.[21] Consumption of nitrates from contaminated well water is associated with methemoglobenemia ("blue baby syndrome") in infants.[22]

The chain of events leading to such water pollution typically goes something like this: Nitrogen-containing fertilizer in a solid, liquid, or gaseous form is applied to a corn crop; typically, half the nitrogen in the fertilizer is taken up by the plants and the other half escapes, mostly in nitrate form, ending up in streams or ground water or vanishing into thin air; the harvested corn goes to a feedlot, where a steer turns each ten pounds of corn into one pound of meat, retaining part of the corn's store of nitrogen and ridding itself of the rest via urine and feces; some of the meat ends up roasting in an oven in an American suburb; and if, as is likely, the people to whom it's served have already gotten their daily requirement of protein before ever sitting down at the dinner table, much of the nitrogen they ingest from the roast ends up in the city's sewer system within a few hours. One study found that, typically, less than 15 percent of synthetic nitrogen that's manufactured actually ends up nourishing humans; when grain is fed to cattle for meat production, the nitrogen incorporated into our bodies amounts to less than 5 percent (see Table 6.2).

In a report prepared for the Ecological Society of America (ESA) in the late 1990s, a group of prominent scientists calculated that through agriculture and industry, humans have doubled the amount of nitrogen that cycles through the terrestrial biosphere.[23] That is, we are taking as much nitrogen from the atmosphere and fixing it into biologically active molecules as nature was fixing, in total, before the dawn of agriculture. Nitrogen is essential to life, but living things evolved in a pre-agricultural world in which there was very little usable nitrogen just lying around; at any one

Table 6.2 Of 100 pounds of synthetic nitrogen fixed in the Haber-Bosch process, most will be lost before it can be retained in the human body. These are the quantities lost along the way, when crop products are either eaten directly or fed to cattle, whose meat is then eaten.[24]

Fate of nitrogen	Pounds of synthetic nitrogen (out of 100 pounds manufactured)	
	Vegetarian diet	Carnivorous diet
Is lost in the factory, on the way to the farm, or during application	6	6
Escapes the crop's roots into the air and water	47	47
Is lost in food production and processing	5	24
Is eliminated in consumer's urine and feces	12	3
Nourishes the consumer	14	4

time, most of it was locked up in living organisms. The flood of nitrogen unleashed on the Earth by industrial fertilization and the growing of leguminous crops has played havoc with well-established systems, the shock being analogous to that of newly discovered gold or oil in a previously impoverished country. Among the consequences cited by the ESA study were:

- Acidification of soils, which damages plant growth creating deficiencies of calcium and potassium and excesses of toxic aluminum.
- Contamination of drinking water with nitrates, which are hazardous to human health.
- Smog and acid rain.
- Pollution of estuaries and coastal waters, which has reached the point of creating scores of so-called "dead zones" in which for several months each year there is insufficient oxygen to support any marine life at all.
- Steep losses of biological diversity in many ecosystems.

And now we're upping the ante, embarking on the most wasteful nitrogen-and-energy spending spree of all: turning crops into ethanol for vehicle fuel. Corn that is used to make ethanol will be raised just as it is for cattle: by eroding soil, burning vast quantities of fossil fuels, and flushing ever more nitrogen into the rivers and oceans. Perennial grasses used to produce cellulosic biofuels will be less wasteful than corn but will still have to be heavily fertilized with nitrogen to get large yields. Ethanol can feed only a very small part of the huge and growing demand for liquid fuels, and at an enormous cost.[25] But the impact on the land will be multiplied as nations look to the tropics and their highly productive climates for maximum biomass production. That conversion to fuel cropping could treat millions of malnourished families to lovely views of green but foodless landscapes.

In competition for energy, synthetic nitrogen, and land, ethanol will surely go straight to the head of the line, because the automobile culture always gets what it wants. Sure enough, as of spring 2007, financial analysts were advising investors that natural-gas stocks were a good buy, because booming ethanol production would mean more consumption of gas in making fertilizer for fuel crops (and cooking the biofuels as well).[26]

NEEDS AND WANTS

Whenever the global effects of excess nitrogen are mapped, the damage is clearly concentrated in and around the wealthy countries of the West. Big industry, including industrial agriculture, can afford the energy needed to churn out nitrogen compounds at a rate far beyond what society can use efficiently or dispose of properly. Meanwhile, the people left behind in poor nations scramble for chronically short and costly supplies of both fuel and nitrogen.

Ask someone whose children's lives depend on the industry that pulls nitrogen out of the air and gets it into food crops, and she'll probably tell you there's no higher use for natural gas. But in affluent societies that take food for granted, gas (in the industry's words, "one of the cleanest, safest and most useful of all energy sources"[27]) can provide a lot of options that, after a while, start looking like necessities: keeping the house comfy on an August afternoon, dining on corn-fed barbecue, taking an ethanol-fueled SUV to the store to pick up more organic milk, or letting the backyard jacuzzi soak away the daily stress that can build up in a world gone haywire.

Without a right to food, people have no rights at all. So when there's a worldwide rush on a mineral resource essential to the production of adequate food—when reliance on markets is a problem, not a solution—non-market measures are needed to ensure that farmers are free to raise essential food crops. The Food and Agriculture Organization (FAO) of the United Nations has nonbinding "Right to Food" guidelines stating in part that:

> States should consider specific national policies, legal instruments, and supporting mechanisms to protect ecological stability and the carrying capacity of ecosystems, to insure the possibility for sustained, increased food production in present and future generations, prevent water pollution, protect the fertility of the soil, and promote the sustainable management of fisheries and forestry.[28]

A firm legal basis for ensuring that all people have access to the means of food production is the UN's 1976 International Covenant on Economic, Social and Cultural Rights, which recognizes "the right of everyone to be free from hunger." The treaty has been ratified by more than 150 nations. The United States is not among them.

Americans cannot expect to support a universal right to food via the roundabout and futile practice of importing natural gas and fertilizer, using them to produce surplus grain, and then exporting the excess to hungry people in countries that could have fed themselves given sufficient resources. Every nation must have the means to grow its own food sustainably, with efficient recycling of crop, livestock, and human wastes. And when those nutrients aren't sufficient—and that, unfortunately, will be true in nations with high population-to-land ratios, even ones with effective family planning—farmers need guaranteed access to supplemental fossil fuels and fertilizers as well.

Mainstream economists are ready with the usual easy prescription: as the price of natural gas goes up, they'll tell you, people and nations will become more serious about conservation. But, as economist Herman Daly argues, mainstream economic theory is handicapped by its inability to differentiate between "absolute wants"—such as food, water, shelter, and clothing sufficient to provide a decent life—and "relative wants"—which grow out of a desire to acquire more and better goods and services, relative to the norm of the society. Absolute wants can be quantified and limited, but relative wants are infinite, because the more successfully they are fulfilled, the more additional desires they create. Daly observed that economics has its roots in an era when most wants were absolute—based on fundamental needs— yet scarcity was relative; that is, when one natural resource fell into short supply, another was soon found to serve as a substitute. But, he argues, by the late twentieth century, we'd entered an era of absolute scarcity, facing finite supplies of irreplaceable resources and the destruction of entire ecosystems on the only planet we have. At the same time, economies had evolved to satisfy relative wants on a grand scale while neglecting absolute wants throughout the world.[29]

The combination of relative wants and absolute scarcity, argues Daly, is not only leading to suicidal growth but also deepening the misery of the planet's have-nots. An economy that's designed to satisfy everyone's relative wants requires both infinite growth (because such wants are unlimited) and perpetual inequality (because in a more equitable world, no one's consumption could be higher than that of others—it wouldn't be relatively "better").

When it's keeping people warm in Moscow or well fed in Manila, natural gas is satisfying absolute wants. But the global market doesn't necessarily favor those critical needs over the pursuit of ever-expanding relative wants. To make matters worse, natural gas is just so darn handy, and it's not always easy to sort out its worthy and unworthy uses. Latched onto increasingly as a more benign substitute for other fossil fuels, it's playing the role of methadone in humanity's vain attempt to ease its withdrawal from coal and oil. And market forces tend to go haywire when dealing with addictive substances.

Nitrogen fertilizer produced largely with natural gas made it possible for us to overpopulate the Earth, and now we're hooked on it. Someday, as reserves of fossil fuels dwindle, our descendants will come to inhabit a less crowded planet, on crops fed entirely by sunlight and natural fertility. Whether that population decline happens humanely through planning and restraint or cruelly through catastrophe depends largely on how we design our economic system to manage nonrenewable resources. For now, in rapidly industrializing nations that attempt to emulate the West's rise to economic power, those resources are being managed very badly indeed. The next chapter examines one frightening consequence that's still just a prediction but may soon be a reality.

7

DOWN-TO-A-TRICKLE ECONOMICS

With fuel and fire, then, almost anything is easy.
William Stanley Jevons, *The Coal Question*, 1865[1]

In the 1970s, TV viewers across the world were horrified by images of skeletal famine victims in the Sahel region of Africa. By the time consistent rains returned to the region years later, a million people had starved to death. Some said that too many people were trying to live in a region too dry to support them. Some said lack of development in the Sahel's economy, especially its agriculture, made it vulnerable. But there was another factor, unrecognized until recently: Western industry had contributed directly to the tragedy. Research now shows that during that time, huge clouds laden with sulfates and other pollutants drifted from Europe and North America over the North Atlantic, shading and cooling the ocean surface and partially shutting down the weather systems that normally send rain clouds over the Sahel each summer.[2]

In the years since, the old industrial powers, while failing to curb emissions of globe-warming carbon dioxide, have managed to cut back on their emissions of the pollutants that dried out the Sahel. But now a similar stew of pollution may imperil food production in India and the other nations of South Asia. This time, Western industry isn't the source, at least not directly. But by no means can we claim to be innocent bystanders.

DIMMING, GLOBAL AND LOCAL

Sunlight intensity, averaged across hundreds of locations on all continents, decreased by 1.5 to 3 percent per decade from the

1950s to 1990s. The dulling of the sky can be traced largely to the burning of fuels that released sulfates, nitrates, and old-fashioned soot. Those and other pollutants absorb or reflect a portion of the sunlight that normally would reach the Earth's surface.[3]

When they were first being reported at the end of the twentieth century, these findings were controversial; however, subsequent research has helped confirm the reality of what's called "global dimming." And as it turns out, to borrow the old saying about politics, all global dimming is local. Global-warming carbon dioxide, when released by burning of fossil fuels, mixes throughout the atmosphere and stays there until scavenged by green plants or absorbed by the oceans. But global-dimming pollutants float for a relatively short time, eventually falling back to earth or sea. They drift but don't spread evenly across the globe, and if they aren't replenished regularly, they clear out and the skies brighten.

The deepest dimming during the past half-century occurred in the Northern Hemisphere, most intensely in the most heavily populated regions, and especially in the United States, with its voracious energy consumption. Anti-pollution efforts in the industrialized West, along with the 1990s economic crash in the former Soviet Union and Eastern Europe and the large-scale relocation of manufacturing to Asia, curbed the release of pollutants (other than carbon dioxide) in the West, and that appears to have led to overall global brightening in the last decade or so. Some analysts now say the sudden onslaught of hot years over much of the North since 1990 actually represents a longer, more gradual warming trend that was masked back in the 1960s, '70s, and '80s by a shady layer of soot and sulfates.[4]

Now, food production in rapidly industrializing South Asia is imperiled by mile-and-a-half-thick leviathans known as "atmospheric brown clouds." The subcontinent and northern Indian Ocean have seen continued darkening, associated with the emergence of extensive, murky clouds. Fed increasingly by combustion of coal, diesel, and gasoline, brown clouds have been returning, darker and larger each winter, over South Asia and the northern Indian Ocean. They've cut the amount of sunlight reaching the land and ocean surfaces by approximately

8 percent between 1930 and 2000. While shading and thereby cooling the surface, the brown clouds absorb heat and warm the atmospheric layer in which they hover. That has several nasty consequences: reduced evaporation from the ocean surface (which means less moisture available for rain); warmer-than-normal clouds that contain more fine particles of pollution and can hold more moisture without releasing it as rain; and perhaps most ominously, a potential weakening of the climatic engine that drives the monsoon rains.[5] That could mean more crop failures across much of India, Pakistan, and Bangladesh, and it could tip already drought-afflicted areas like south India's Anantapur district into ecological and humanitarian crisis.

At the southern end of the state of Andhra Pradesh, Anantapur may provide a grim preview of South Asia's future. Lying on a stretch of the Deccan Plateau between Hyderabad and Bangalore— the country's two high-tech, traffic-choked foreign-investment capitals—this impoverished rural region never sees a very good monsoon. It lies in a "rain shadow" from India's southeastern mountains, and as a result, its average annual rainfall is only about 20 inches, often concentrated in a few downpours that hit sporadically between June and September. And even that meager monsoon is increasingly undependable: Of the nine years since 1930 that saw rainfall below 16 inches, six have occurred since 1980 and two since 2002.

A recent study by India's National Climate Center showed that over the past century, 12 of 36 regions in India, including the region that includes Anantapur, have seen decreasing annual rainfall.[6] But despite living in the driest part of southern India, the 3.6 million people of Anantapur district continue to rely on agriculture as the foundation of their economy, indeed their survival. Now, dimming clouds spewed out by the booming, mostly urban, demand for electric appliances, automobiles, and other fossil-fuel-hungry features of twenty-first-century Indian life could undermine Anantapur's survival in ways that centuries of persistent "natural" droughts have not.

Traditionally, Anantapur's farmers have dealt with their bad draw in the climatic lottery by growing tough subsistence crops:

pearl millet, finger millet, deep-rooted legumes like pigeonpea, and, on better soils, chickpea. Over the past two decades, cash-crop peanut mania swept the district, eventually covering its arable land in a near-monoculture. But drought, soil exhaustion, and a plant virus have driven peanut yields down and reduced the harvested seed's oil content from almost 50 percent down to 36 percent. Because the crop is grown mainly for cooking-oil production, farmers are getting lower prices for smaller crops.[7]

Fully 80 percent of the district's rural people are small farmers, not fat-cat landlords or landless laborers, but that relatively well-balanced farm economy is getting harder to maintain. Economic pressures, coming on the heels of increasingly erratic rainfall and depletion of groundwater supplies, have earned Anantapur the dubious distinction of being a top district for farmer suicides, which reportedly number in the thousands. The water table has dropped as much as 15 feet in some places, and more wells are being drilled ever-more deeply to get at ever-less water. It's also reported that 10 to 15 percent of farmers have fled the crisis. Y.V. Malla Reddy, director of the Ecology Center run by the development non-profit Acción Fraterna/Rural Development Trust in Anantapur, told me there's no starvation yet in the district, only "slow poisoning." He said: "We are suffering from 'distress migration,' when economic crisis drives people off to the cities, as well as 'brain-drain' migration, when the more well-off and better-educated young people are drawn away by the big-city salaries they hear about."

In the last week of 2006, Reddy took me to visit farmlands around the village of Velikonda, one of dozens of watersheds in the district where farmers, assisted by Acción Fraterna, are using water-harvesting methods, a more diverse array of crops, and natural pest control in an effort to sustain their communities and food supplies over the long haul. It's clear that Velikonda and some other villages in the district, accustomed as they are to surviving hard times, are not going to give up without a fight. With the help of engineers, hydrologists, agronomists, and local laborers, and organizing themselves into teams of 15 farm families (of various low castes and non-castes, none of them well-to-do), people in Velikonda and a host of other villages are planning and building large water-conservation networks. Using mostly hand labor, they have built thick, chest-high earthen berms around the downhill edges of fields to trap precious rainwater that would otherwise run off into gullies during storms. More than 100,000 acres in the district are now protected by such berms.

They are hand-digging deep ponds as well, to hold rainwater that they carry to new orchards of mango or custard-apple trees. Farmers pay a percentage of the cost of berms, drainage outlets, and ponds on a sliding scale, with the rest coming from Acción Fraterna and the government. They are also getting out of the old economy in which they sold peanuts in order to buy nutrient-poor, government-subsidized rice. In an effort led largely by women, they are re-diversifying their cropping system with nutritious crops they can both consume and sell: millets, sorghum, pigeonpea, broad beans, cluster beans, chilis, coriander, and many more. They have stopped buying costly pesticides, turning instead to natural products like neem seed extract. They are growing non-crop plants like milkweed to trap insects instead of trapping themselves in debt to buy chemicals. They are growing large leguminous plant species on the water-holding berms, to be cut and spread back on the land to add organic matter and nutrients; as we have seen, places like South India cannot wean themselves completely from synthetic nitrogen fertilizers, but such practices can reduce that reliance. Where these self-organized community efforts have taken root, they have beaten back the individual despair that had developed under the brutal logic of the national and international economy. The work is on a colossal scale and no doubt exhausting. Despite, or maybe because of that, the atmosphere in the villages is electric.

DARK HORIZON

Yet even if the people of Velikonda and thousands of other villages make every right move within their local, water-limited means, the global economy may not be finished with them. India's industrial transformation has added 50 percent on top of the pollutant emissions that have come with population growth since 1930.[8] That foul output has accelerated with the opening of the nation's economy over the past decade and a half. India's integration into world markets has meant bigger profits for multinational corporations and cheaper shirts and software for Western consumers. It has also swelled those brown clouds that could choke off the monsoon.

Manufacturing of consumer goods for the domestic market and for export produces pollution, both directly and by increasing residential demand for home electricity and cars. From 1990 to 2002, India's population grew by 24 percent while its electricity consumption leaped by 105 percent.[9] Well over half of that power is currently generated with coal.[10] With natural gas becoming more costly (and, as we've seen, more important in food production) and with rising grassroots opposition to flooding of precious farmland for hydroelectric projects, those "cleaner" sources will probably be giving way to even more coal use in future years. At the same time, every day of the year, thousands of newly built vehicles, many running on dirty diesel, roll out to join India's near-impenetrable traffic jams.

India's breakneck growth rate of recent years was catalyzed by $57 billion in foreign investment since 1991. Of that total, $38 billion has come in since 2000.[11] The US is the leading destination for Indian exports and the second-biggest source of foreign investment in India.[12] (The biggest is the tiny island nation of Mauritius, which is really just serving as a way-station for money coming from tax-averse investors in the US and other countries.) Investment by expatriate Indians in the Mumbai stock market increased 200-fold from 1996 to 2006.[13]

General Electric, Microsoft, Coca-Cola, Pepsi, Ford, Monsanto, Hughes Electronics, Fluor, Whirlpool, and a host of other US corporations that invest in, outsource to, import from, and sell all over India are cheering the country's growth not from the sidelines but from the playing field. Software and telecom get all the headlines, but about two-thirds of foreign investment is in dirtier activity: transportation, electrical equipment, fuels, chemicals, drugs, cement, and metals.[14] A statistical study of data from the world's "less developed countries," published in 2003, concluded, "The central finding of these analyses is that *dependence on foreign capital accelerates* the rate of growth of CO_2 emissions in less developed countries" (their italics).[15] That study did not examine other pollutants, but in India, the impact of the foreign-investment-induced boom is striking. Investors have every incentive to overlook the grimy clouds that are being belched out by power plants, trucks, and cars, because India's economy is growing the same way the powerhouse economies of the West grew: with products and technologies that burn every ton or barrel of fossil fuel that comes within their reach. The boom is (1) sickening rich and poor alike in the cities; (2) creating brown clouds that threaten to tip rural areas from chronic hardship into drought-stricken crisis; (3) helping undermine both itself and the world economy by helping make global warming irreversible; and (4) bestowing its benefits on a relative few.

Let's look more closely at point (4). Both statistics and the naked eye confirm that the gap between scattered pockets of prosperity and vast expanses of deprivation is growing. Steadily increasing economic openness has been accompanied by steady growth in standard measures of inequality. The gap between wealthier cities and poorer districts like Anantapur has also grown.[16] The new economy's fruits may not have been shared widely, but its environmental ravages already have been. India ranks a sorry 126th among 177 countries for a United Nations "human development index." The index, says the UN, measures people's chances of achieving a long, healthy life and decent standard of living. India is now in a lower percentile than it occupied in pre-boom 1990.[17]

Brown clouds have become a major obstacle to a long, healthy life in India's cities as well as the countryside. Respiratory problems, for example, are on the rise.[18] The big boom has made state-of-the-art medical care much more widely available in cities for those who can afford it. But most villages have no functioning health care facility anywhere close by, and most villagers cannot afford the care offered in the cities.[19] As we saw in Chapter 3, poor Indians' health is also being sacrificed to produce drugs for fighting other people's diseases.

It could have been different in India. From the 1950s through the 1980s, the country held Western consumerism at bay, keeping open the possibility for a new kind of development that is not synonymous with growth. The subcontinent is woven throughout with a reliable passenger railway system—the planet's largest. India ranks third in the world for solar power and fifth for wind power. Those and other such strengths could have been built upon. Instead, the new, global India is steering its economy down that same smoggy road to wealth that was traveled so enthusiastically by the twentieth-century industrial powers. But this time it may be a highway to hunger.

"Trickle-down" economics never works as advertised; indeed, for the poor, its results are usually worse than those that grow out of simple neglect. The form it has taken in India might be called "down-to-a-trickle" economics. If the boom continues to darken and thicken the atmospheric brown clouds that, in turn, are threatening to disrupt the monsoon, it could destroy everything that the resolute farm communities of Anantapur, and rural people across the subcontinent, have managed to accomplish. Computer models predict that nationwide monsoon droughts, which historically occur an average of two to three years per decade, could rise to as many as six years per decade under the influence of brown clouds.[20] If Anantapur is affected as badly as the nation as a whole (and the models appear very uncertain about local variations), agriculture might just become impossible. Because the district lies in a rain shadow, farmers say they already count on drought at least six years in ten; brown clouds conceivably could make that a perfect ten out of ten.

No sudden, human-made climatic change in a random direction has, as one might expect, an equal chance of being either harmful or beneficial. Because life on Earth evolves toward equilibrium with its current environment—and on a long time scale—and because industrial civilization has become so complicated, fragile, and vulnerable, any rapid climatic change, including that which brown clouds may cause, is almost guaranteed to prove a disaster with no silver lining.

No one knows, for example, how the complex tug-of-war between global warming and local dimming will turn out. But the results for South Asia are unlikely to be pleasant. The leader of the Atmospheric Brown Clouds Project, Dr. V. Ramanathan of the Scripps Institution of Oceanography, has said: "Some years the aerosols [i.e., the pollutants causing dimming] might win and in some years the greenhouse effect may win. So we are concerned that in coming decades the variability between the two will become large and it will be difficult to cope with rapid changes from year to year."[21]

THE 66,000-LB GORILLA IN THE LIVING ROOM

Were India passing through an inevitable, sooty phase of development, a dark industrial tunnel that would open out eventually onto a green, prosperous future for all, then we could cheer them on: "OK, great, keep it rolling along and get through this unpleasantness as quickly as possible." But while Western nations of the late nineteenth and early-to-mid twentieth centuries were growing out into an "empty world," ecologically speaking, India is trying the same trick, but finding that it's now a "full world."[22] Looming climatic changes mean that the old growth strategies won't work for any nation, rich or poor.

The phenomena of global warming and local dimming are very different in their origins and their behavior. But they call for the same straightforward corrective: to make deep cuts in energy consumption planet-wide. The rapidly industrializing nations of the South will have to find their own ways to get the energy they need without ecological devastation. But with the average American using 10 times as much energy as the average person in China, and 24 times as much as the average Indian, it's the clear duty of the United States and fellow energy gluttons to take the lead in slashing consumption.

But restraint continues to be trumped by market forces. In the international establishment's view, the failures of the global economy are always in the past, its successes always waiting just around the bend in the road ahead. A 2006 Asian Development Bank report on India asserted: "Free trade alleviates poverty, if at all, through its growth impact," while at the same time conceding that the country's trade-supercharged growth had increased income inequality.[23] On balance, the report concluded, free-trade policies should not be abandoned just because they disproportionately enrich the already more well-to-do sectors of the economy. Better to let the "haves" of the world have more, because diversion of some of those riches, "if at all," might help reduce poverty. There's little doubt that Bank officials would respond affirmatively if asked, "Are you working for sustainable development?" Since first being formulated in the 1987 UN document *Our Common Future* (a.k.a. the "Brundtland Report"), "sustainable development" has become a refuge for those who yearn to do something about poverty and ecological devastation without addressing the fundamental contradiction between growth and real sustainability. As one keen observer concluded: "In fact, sustainable development is code for 'perpetual growth'."[24]

Clearly, the planet would benefit both from having fewer people and from reduced ecological impact per person. But when representatives of poor and rich nations meet to decide how to do that, they generally come to a standoff that ends something like this: "You've got too many cars!" "Oh yeah? Well, you've got too many people—and you *want* too many cars!"

History shows that the typical way any nation stabilizes its population is to raise its economic standard of living—a process social scientists call the "demographic transition." The transition, which is also usually associated with urbanization, works partly because it costs parents a lot more resources to raise a child in a rich country. The transition depends not on the rise in incomes for individual families, but on the overall level of consumption in the country where they live. So rising inequality in countries like India is no problem, is it? As the country as a whole gets richer, population pressures will recede and there will be more of everything to go around. Right? No, because India and other would-be economic powers are trying to grow into an already full world. Brown clouds are among the blowback effects of doing that. To cripple one's food and water supplies in the name of development and population stability is clearly self-defeating.

It's not a fair fight, to be sure. The West has already staked out most of the resource territory available for demographic transition. The United States, the United Kingdom, and Australia together consume more energy each year than India, Pakistan, Japan, China, the Koreas, Southeast Asia, and the Pacific Islands combined. Those Asian-Pacific nations have more than 10 times as many people as we do in the US, UK, and Australia, and we use almost 10 times as much energy per head as they do.[25]

A 2003 study by Melanie Moses and James Brown in the journal *Ecology Letters*[26] sharpens the horns of this dilemma and reveals what really lies behind "sustainable development." They analyzed data from more than 100 countries and concluded that humans are far from unique among mammals. In fact, we obey a general biological law: The greater the energy consumption by individual animals of a species, the fewer offspring they will produce and raise. From little monkeys to big apes to prehistoric humans to subsistence farmers to commuters in their SUVs, increases in energy consumption lead to smaller families. (For you math fans, the decline in fertility is proportional to the cube root of per-animal energy consumption.) A blue whale needs a much bigger vascular system and a lot more energy than does a rabbit to deliver nutrients and oxygen throughout its body. An American toddler, in turn, is hooked up to a support system—a planet-wide industrial infrastructure—that dwarfs the blue whale's wholly biological support system. But it's not a straight-line relationship. Larger organisms get less benefit from each additional kilocalorie than do smaller ones; science has shown that to supply bigger organisms with bigger quantities of resources requires systems of greater complexity that have to reach over greater distances.

Humans have taken matters to the extreme with our unique ability to extend "energy networks" far beyond our physical bodies. (Moses and Brown note, "The per capita energy consumption rate in the United States is ... the estimated rate of energy consumption of a 30,000-kg [66,000-lb] primate." With human proportions, such a creature would be over 100 feet tall.) As we've drawn upon greater quantities of fossil fuels and other resources, we have built societies in which people have education, contraceptives, and pension plans, all of which encourage smaller families. The people of rich nations might like to believe that high consumption has thereby freed them from the laws of nature. But Moses and Brown's analysis says that lunch still isn't free: "We hypothesize that parents face a tradeoff between the number of offspring and the energetic investment in each offspring ... the perceived energetic investment (including material goods and education) required for a child to be competitive in a given society is greater in more consumptive societies." Raising a creature who's the resource-depleting equivalent of a 66,000-lb gorilla isn't cheap or easy. And, as with other animals, each additional gain in reproductive restraint or other biological characteristic requires a greater quantity of energy than the one before.

India is not flush with deposits of mineral resources. The well-frayed cliché tells us that such a nation must rely instead on its vast "human resources." In the current context, the real meaning of "human resources" is "large numbers of English-speaking people who will create far more wealth than will be returned to them in their paychecks." But will that be enough to fuel a demographic transition to the prosperous, low-birth-rate future that India seeks? According to Herman Daly, standard economics equations that combine labor and resources lead to absurd conclusions, for example that "we could produce a 100-pound cake with only a pound of sugar, flour, eggs, etc., if only we had enough cooks stirring hard in big pans and baking in a big enough oven!"[27] The other thing India has to offer up to the world for sacrifice is its landscape in places like the industrial estates of Patancheru or the parched fields of Anantapur district.

Were India's whole population to follow the American path of resource use, the ecological impact would be like doubling the planet's human population from 6.6 billion to 13.2 billion.[28] If we expect India and other global have-nots to break out of Moses and Brown's mathematical model, the US and other rich countries will have to break out first. Instead, we continue to enjoy the high-energy life, and our economies goad those of less wealthy nations to follow our example. Then we carp about their having too many people while we take full advantage of the cheap labor power that those "too many people" can provide.

India's population is 70 to 75 percent rural, and hunger is commonplace in both cities and villages. So it's no surprise that food production is a hotter topic in India than in the West. Even in the era of software and call centers, farmers' irrigation pumps get preference over city-dwellers' refrigerators when electricity becomes scarce. In south Indian villages and slums, people passing on the street or path will often exchange a not-so-rhetorical greeting: "Hi—have you eaten?" But the greater the influence of software development, communications, pharmaceuticals, and other Western-oriented, high-profile constituents of the economic boom, the greater the pressure on agriculture will become. The most absolute of absolute wants is for food and water, and there's not a lot of room there for cutting energy. The cuts have to come in the satisfaction of relative wants. That means big reductions in the West and among economic elites wherever they are. Not doing so will put an even bigger burden on those least able to bear it, both directly and through surprises like atmospheric brown clouds.

Of course, in biology, no mathematical relationship is absolute. Looking at those nations that deviate from Moses and Brown's overall trend can be as instructive as studying those that follow it. For example, birth rates in ten oil-producing nations whose citizens have almost unlimited access to fossil fuels are much higher than would be predicted by the energy-fertility equations. Meanwhile, Cuba, when compared with Central America and the larger nations of the Caribbean, has similar per capita energy consumption but only half the birth rate. Cuba's lower rate of population increase is generally attributed to its high degree of economic equality, a rarity in Latin America.

The generally close relationship between energy consumption and fertility decline, however, suggests that humans, like members of all species, tend to keep consuming until, inevitably, they hit a limit. That limit is miserably low if you're living in a thatch shack in Anantapur district; in a Dallas suburb, it's far too high.

All other species live off of what they take into their bodies and have no way to sneak past Moses and Brown's equations. But human societies have both a lot of excess energy consumption waiting to be cut and the knowledge and motive to do it. If we put these hefty brains of ours to good use, we could become the first species to restrain both our numbers and our consumption. With big business throwing up the usual roadblocks, the solution will require more ecological and economic justice, not more enrichment of the investing class.

Instead, scientists and policymakers are focusing on outrageous technological tours de force like carbon sequestration, stratospheric sulfur seeding,[29] and colossal, space-based mirrors.[30] Growth-dependent economies were built around the fossil-fuel power bonanza and have no way to handle the consequences of the deep energy cuts that are necessary. Global capitalism will not, indeed cannot, give up the easy exploitation of concentrated energy that need only be mined or pumped. Some communities in places like Anantapur district, having seen few benefits from the global high-energy economy, have set about wriggling free of it. But they're unlikely to escape brown clouds or other gifts bestowed on them by down-to-a-trickle economics. Unless nations both poor and rich find a way to development that violates Moses and Brown's equations, prospects are looking pretty dim.

The fossil fuels that remain could and should be used to subsidize the building of a society that can thrive without those fuels in the long run. That means giving top priority to greener systems of food production. But if those systems end up exploiting their own workers at the bottom of the wealth scale in order to serve customers at the top of the scale, nothing will have been gained.

8

SUPERNATURAL FOOD

The labourer consumes in a twofold way. While producing he consumes by his labour the means of production, and converts them into products with a higher value than that of the capital advanced. This is his productive consumption. It is at the same time the consumption of his labour-power by the capitalist who bought it. On the other hand, the labourer turns the money paid to him for his labour-power into means of subsistence: this is his individual consumption … The result of the one is that the capitalist lives; of the other that the labourer lives.

Karl Marx, *Capital*[1]

Millions of words—many of them insightful, many others annoyingly rhapsodic—have been published in recent years about natural alternatives to industrial food. Too often they converge on a single issue: the consumer's choice of what to eat. The United States is now a largely urban nation, so battles over the nature of food are being fought most prominently in the retail world. It's through food marketers that most health- and nature-conscious consumers, writers, and activists are trying to exert their influence. But efforts to break natural and organic food out of their upscale niche and get them into the shopping baskets of the American majority continue to founder.

Any economy that's going to "green" itself will have to make all the elements of a greener life accessible to the bulk of the population—people who also would be the producers of those elements. Meanwhile, any business has to take the value generated by its employees and decide how much will go to pay those employees, how much will be retained as profit, and how much will be sacrificed to lower prices to entice consumers. That's hard enough in the usual free-for-all of business competition, but if an aspiring "green" company is determined to pay on top of that all the normally unpaid ecological costs of its actions, it will become that much more difficult to juggle the demands of workers, management, shareholders, and customers.

GOLIATH JUNIOR vs. GOLIATH SENIOR

As you head upstream from the kitchen, through supermarkets and food processors to the farm and ranch, the consumer's grip on the system weakens, while the exploitation of nature looms larger. And the best of intentions risk being subverted. The organic- and natural-food pioneers of the 1960s and 1970s, in the words of one observer, "surely didn't intend for organic to become a luxury item, a high-end lifestyle choice."[2] But so far, throughout much of society, that's exactly what it's been.

Now Americans are being treated to a fascinating spectacle, a grand experiment testing whether capitalism can deliver ecologically sound food to a majority of the population while satisfying its own fundamental need for growth and profit. Two Fortune 500 companies, Wal-Mart (no. 2) and Whole Foods Market, Inc. (no. 449) are attempting to do that, each through its own tried-and-true methods—methods as different as Pop-Tarts and quinoa crackers. The two Goliaths (one 56 times the size of the other) are looking to capture that big stretch of socioeconomic territory that separates them. Wal-Mart, the shopping home of American families who live below the median income, made headlines in May 2006 by announcing a big increase in its marketing of organic food and cotton clothing. By the end of 2006, probably not by complete coincidence, the 196-store chain Whole Foods, leader in the natural-food market and purveyor of edible wonders to the prosperous, announced that its long string of double-digit growth years was ending, with projected 2007 growth dipping to "only" 6 to 8 percent.

Wal-Mart and Whole Foods do have one thing in common: They're both rooted in what might be called the Great Man theory of economics. Wal-Mart's success as the world's largest retailer is attributed in large part, by friend and foe alike, to its legendary founder Sam Walton. Whole Foods has been guided for more than a quarter-century by the charismatic, unabashedly idealistic, and fiercely libertarian John Mackey.

In early 2007, Mackey announced that Whole Foods would accelerate growth by the tried-and-true method of buying out the biggest competitor. But its planned acquisition of Wild Oats Marketplace, operator of more than 70 stores in the US, was blocked a few months later by a Federal Trade Commission (FTC) lawsuit. Now embroiled in antitrust litigation, natural food had truly arrived as an economic force. In August, with the FTC suit still unresolved, Mackey helped prove the government's case with an email to his board. In the message, he wrote that his target Wild Oats "is the only existing company that has the brand and number of stores to be a meaningful springboard for another player to get into this space ... Eliminating them means eliminating this threat forever, or almost forever."[3]

To anyone who has wandered through a Whole Foods Market, there's no secret to the company's success. It's gotten where it is by selling food at high prices to people who can afford to buy it. Mackey is frank about that. In 2005, he told the *Independent*: "You can't have it both ways. If you want the highest quality, it costs more. It is like complaining that a BMW is more expensive than a Hyundai. Yes, but you're getting a better car."[4] And that's why you don't see the parking lots around a Whole Foods Market packed with rusty Fords and Plymouths. Of the company's 170 stores in the US in 2006, not a single one was located in a zip code with an average 2003 household income as low as $27,800—the most common Whole Foods hourly-job salary,[5] which in turn, exceeds the annual income of 27 percent of American households. Only 5 percent of its stores were in zip codes with average incomes at or below $43,300, even though half of all US families that year lived on less. Fully 50 percent of the zip codes hosting Whole Foods stores surpassed $72,000 in average 2003 income. Half of *those* exceeded $100,000.

In her book *Why People Buy Things They Don't Need: Understanding and Predicting Consumer Behavior*, luxury marketing guru Pamela Danziger advises companies on how to lure people into spending disposable income on high-priced products: "Some consumers gain satisfaction from finding a shopping fantasy they can act out. For others, it is the power they feel from finding something and being able to buy it. Interestingly, the consumer's feelings often have more to do with the act of purchasing than with the object that the shopper buys."[6] Feature articles about Whole Foods inevitably pile on the adjectives when describing the beauty and presentation of the natural—almost supernatural—merchandise in its stores; probably the most entertaining of this genre was the 2005 *Forbes* magazine piece entitled "Food Porn," which concluded that the company "thrives by presenting food as theater, playing up the pious organic angle even as it peddles tempting offerings of culinary excess." In the article, a Whole Foods executive's explanation of his company's marketing strategy would have fit perfectly in Danziger's book on luxury buying. He said: "More than half of shopping decisions are made on impulse. When you shop, we engage your senses. We want to romance the food."[7]

Wal-Mart is now threatening to take organic food's image way, way downscale, where romantic impulses are overruled by thin wallets. In a *Wall Street Journal* interview piece entitled "Whole Foods' CEO intends to stop growth slippage," Mackey was asked how organic and natural food can become the standard rather than the exception. He responded, "It'd have to be cheaper. Most people shop based on price. ... There are others who value quality more, but they're in a relative minority."[8] His new rival, Wal-Mart CEO Lee Scott, sees his customers' desire for cheap goods in a different way, one that has little to do with whether they "value quality." Scott told his directors and executives in 2005, before the company's organic move, that wage and price increases are out of the question at Wal-Mart, because "our customers simply don't have the money to buy basic necessities between paychecks."[9] I happened to be in Fayetteville, Arkansas on the day of Scott's speech to the company's 2007 shareholder convention. There, his message was simple: "We save people money so they can live better."[10]

Agribusiness takes pride in the fact that Americans now spend only 10 percent of their income on food—half the share they spent in 1920. That 10 percent is an average, and like many averages, it obscures more than it reveals. Low-wage earners may have to spend 25 cents of every dollar they earn on food, even cheap Wal-Mart food. But the top executives at either Whole Foods or Wal-Mart could easily live on Whole Foods' sumptuous cuisine for less than one tenth of 1 percent of their salaries-plus-bonuses.

Were consumers paying the full costs of food—if the massive ecological damage were halted, if migrant fruit-and-vegetable pickers had a decent wage and quality of life, if family farms and communities were being kept viable, if the military costs of maintaining access to Persian Gulf oil were counted—food would inexorably grow more costly. If Whole Foods, Wal-Mart, and other corporations are going to bring even marginally more ecologically sound food products to the mass market, while at the same time sustaining or increasing their profits and selling more cheaply, the money to do it will have to come from somewhere. Or most likely from some*one*: the worker who produces, processes, stocks, and sells the food.

SHOP WHERE YOU WORK?

Pioneer automaker Henry Ford was much celebrated for announcing his ambition to manufacture cars cheaply enough and pay his employees well enough that they could afford to buy the cars they built. That kind of capitalism-as-populism was widely credited with helping create the great American middle class, but it has since fallen into disrepute, rightly regarded by today's corporate moguls as no more than a sentimental fantasy in a cutthroat global economy. Indeed, Ford's goal could not be generalized to the whole economy; as Marx emphasized, if employers didn't squeeze more value out of their workers than workers require to maintain themselves and their families, the whole system would grind to a halt.

At least in the industries that supply food and other fundamental necessities of life, something might be learned by asking how close employees can come to affording the cost of the goods their own employers sell. In a 2003 article for AlterNet.org, I wrote about my own amateur experiment in that area. Following Ford's example, I asked a simple question: In view of Wal-Mart's vast range of merchandise and "Always Low Prices," could a family whose breadwinner worked at the Wal-Mart Supercenter in the town where I live—Salina, Kansas—afford to supply its minimum needs by shopping there?[11]

I relied on published studies that computed the cost of an "adequate but austere" life for a family with one adult and two children in Salina. The budget included only the basics: shelter, transportation, food, routine toiletries and medicines, and not much more. At that time, housing and transportation couldn't be bought at Wal-Mart, but almost all other necessities could be.

The bottom line: My hypothetical Wal-Mart cashier could not satisfy such a bare-survival budget even if she worked 40 hours per week, more hours than a typical Wal-Mart workweek. She, like a large percentage of the company's actual employees, would need government assistance to get by. And as you might expect, in trying to keep the family within such a budget, I committed them to an array of foods that were boring, unappealing, and not very nutritious—and produced in ways that eco-conscious customers would prefer not to know about. No organic peaches for this family.

In 2006, I asked Henry Ford's question again, this time in a store where knowing how food is produced is an all-consuming preoccupation. I had my cashier work and buy all the family's groceries at a Whole Foods Market in San Antonio, Texas. I used the same list of foods that I'd used at the Salina Wal-Mart: a minimal, USDA-recommended "low-cost food plan." At Wal-Mart, I had not actually bought any food, but engaged only in simulated shopping—mainly because I had little desire to eat the kind of food I was "buying" for the hypothetical cashier. I simulated my shopping at the San Antonio Whole Foods Market as well, because I didn't want to pay the prices. As at Wal-Mart, I selected the least costly food in each food category and the least costly brand or non-brand of that type. Using those prices, I computed the monthly cost of feeding the cashier's family.[12]

At Salina's Wal-Mart, the food portion of the monthly bill had been $232, plus $16 sales tax. (Wal-Mart gives its employees a 10 percent discount on most products, but not on food.) At Whole Foods, the same basket of food came to nearly two and a half times as much: $564. Texas has no sales tax on food, and Whole Foods employees get a 20 percent discount, bringing the total cost for the San Antonio cashier all the way down to $451. The shopping list that generated that register receipt included none of those famed examples of "culinary excess" that draw customers to Whole Foods and on which its profits depend.

How did the hypothetical cashiers come out shopping where they work? The starting wage for a cashier at Salina's Wal-Mart in 2003 was $6.25; that fell $146 per month short of meeting her family's monthly survival budget (including all expenses, not just food) if she shopped where she worked.

Whole Foods employees in three states told me that a starting cashier's wages in 2005–6 tended to be between $7 and $8, but according to Whole Foods spokesperson Ashley Hawkins, a poll of all company regions showed a starting wage at the time of $8 to $10.[13] Assuming, conservatively, that the cost of nonfood necessities in the San Antonio of 2006 was similar to Salina circa 2003, a $10-per-hour employee determined to shop at the Whole Foods there could indeed manage to do so, with the generous employee discount. An $8-per-hour employee could just meet the bare-survival food budget, but with nothing left over. At $7, she would miss her mark by more than the Wal-Mart cashier-shopper. The situation would be worse were she in a state like Kansas that taxes food sales.

Hawkins told me that Whole Foods' full-time turnover rate is 24.7 percent, so the above analyses would apply to approximately one-fourth of employees. She told me the company-wide average wage is $15, and that health care, 401(k), stock option, and stock purchase plans (after about 10–12 months' employment) have helped earn Whole Foods a place on Fortune's list of the "100 best companies to work for" year after year. In return for generating phenomenal profits, the average Whole Foods employee would indeed earn enough to shop at Whole Foods, but only by spending almost 20 percent of her income on food, as compared with the national average of 10 percent. (If the Wal-Mart cashier shopped at Whole Foods, she'd spend more than half her income on food.) In a country where top managers can earn hundreds of times as much as workers, Whole Foods had won praise on the left for capping its pre-2006 salaries at no more than 14 times the pay earned by the average full-time "team member." In November 2006, that ratio was abruptly raised to 19 times the average. The announcement came at the time the company was projecting a big slowdown in growth and a day before its stock value dropped 23 percent. It wasn't clear whether the new 19-to-1 executive-to-worker-pay ratio portended higher executive salaries, lower workers' wages, or both. But one thing was clear: It wouldn't apply to Mackey. He announced that, "following [his] heart," he was having his annual salary slashed to $1. Asked how much that decision had to do with his company's slowed growth, he said, "None. Zero."[14]

All of the buzz over Whole Foods' salary ratio usually buzzes right over the fact that it doesn't cover bonuses and stock options. Mackey earned a salary of $436,000 in 2005—precisely 14 times the average pay, as the policy prescribes. But he made more than $2.2 million on stock-option deals. Four other top executives earned salaries almost identical to Mackey's, and among them they exercised almost $5 million in stock options.[15]

Jeremy Plague was in a group of Madison, Wisconsin Whole Foods employees who managed to form a union local—the only one in the company's history—for a period between 2002 and 2004. The union eventually succumbed to Whole Foods' famously anti-union policies. (Regarding unions, Mackey once told the *Wall Street Journal*: "There's a lot of harm they can inflict by lowering the cooperation in the store, which is ultimately what pays the wages." On another occasion, in a less gracious mood, he compared unions to the herpes virus.[16])

Plague told me: "In my experience, most people really liked working at Whole Foods for the first few months, blinded by the uniqueness of the store and by their hippie rhetoric of how we all mattered. But then people hit the six-month wall where they realize that it's all a bunch of BS, and Whole Foods is just like every other money-hungry corporation. ... It's all a great, worker- and environment-friendly system until you get to the actual people working in the stores, stocking the shelves and ringing you up at the register."[17] A middle-aged employee I spoke with in the San Antonio store told me that she'd advise a young person to apply for work at Whole Foods for the good atmosphere and benefits, but not for the pay.

Each year, the average Whole Foods Market earns $900 profit per square foot,[18] double the food-retail industry's average. And each Whole Foods worker is a powerful wealth generator for the company. The conventional grocery industry (almost one-third of it now in the hands of Wal-Mart) earned about $2,000 profit per employee in 2005–6.[19] In that same year, Whole Foods raked in more than $5,600 profit per employee.[20]

INDUSTRIAL-STRENGTH ORGANIC

The most prominent sector of the natural-food world is organic agriculture, which bans synthetic chemicals and engineered genes and, until recently, held out the vision of a harmonious, human-scale rural society. But since 2002, when the USDA gave "organic" a legal definition by issuing official standards, what had been a trend toward industrialization has turned into a stampede. The standards don't limit the size of the farms or corporations that raise organic crops or livestock. They don't fully protect farm workers or animals. And they don't prescribe how far organic food may be shipped or, if it's imported, how the originating country must treat its farmers and workers.

One result has been an extensively documented industrialization of organic agriculture. Critics have jabbed fingers at large retailers like Whole Foods and Wild Oats that, simply for reasons of scale, often have to seek out suppliers who can deliver massive loads of a relatively uniform product, even out of season. Those companies unintentionally paved the organic highway that Wal-Mart is now traveling, and, too late, fear is spreading. For one thing, everyone's afraid a big share of Wal-Mart's organic products will come from the company's favorite source for just about all other merchandise: China. One prominent critic, writer Michael Pollan, has asked: "How are you going to sell organic food for only 10 percent above irresponsibly priced food [as Wal-Mart plans to do]? It may mean that they'll have to import food from China or other countries. They'll have to buy only from the biggest suppliers."[21]

Coaxing higher prices out of customers and keeping employees from moving on to higher-paying jobs is much easier when all concerned feel that they're connected to a movement that's making the world a better place. But the more that Whole Foods and other retailers grow, the harder it will be to extract premium prices and convince workers that they're changing the world. Critics already charge that corporations are exploiting organic agriculture's feel-good image even when they're selling conventional products. Strolling through the produce section of a newly opened Whole Foods store in Manhattan's Time Warner Center, writer Field Maloney spotted pictures and folksy profiles of neighborly food-growers (like "a sandy-haired organic leek farmer named Dave") positioned above non-organic onions from Oregon and Mexico. Maloney guessed that Whole Foods executives are feeling a little off-balance these days, amidst so much talk about the virtues of locally grown food: "After all, a multinational chain can't promote a 'buy local' philosophy without being self-defeating."[22]

Few exchanges on the subject were more widely followed than a 2006 blog-to-blog conversation between Mackey and Pollan. Mackey started it off by responding to Pollan's sharp criticism of Whole Foods in his 2006 bestseller *The Omnivore's Dilemma*.[23] Pollan wrote back: "I have trouble squaring some of your claims of support for local agriculture with what I see when I shop at Whole Foods. I see more signage about the importance of local produce than I see actual items of local produce."[24]

In another part of the exchange, Mackey implicitly and, I'm sure, unintentionally emphasized his company's role in helping spur the industrialization of organic agriculture: "Without Whole Foods Market's pioneering work and without the growth of our stores and distribution centers, it is very unlikely that the organic foods movement would be where it is today."[25]

If Whole Foods does end up passing the organic leader's baton to Wal-Mart, the eclipse of reality by image could well become total. Within months of Wal-Mart's big organic announcement in 2006, the non-profit Cornucopia Institute was charging the company with placing "organic" signs directly above conventional produce in some of its Wisconsin stores (a violation of USDA's organic standards) and with obtaining its organic food from factory farms and not-so-green foreign sources, including China. Cornucopia notified USDA of the signage problem, but the department was in no hurry to crack down on the nation's largest retailer. Said a USDA spokesperson: "It's not like this is a food safety issue."[26]

Now, with the inevitable slowing of its growth, can Whole Foods keep its investors happy, its elite customers fed a virtuous diet, and its workers paid a living wage? Can Wal-Mart, notorious for putting the price squeeze on its suppliers, really drive the average organic premium down from the currently typical 50 percent to 10 percent? And can the retail behemoth convince its clientele to spend even that extra 10 percent, after decades of proclaiming to America that a price *reduction*, even if it's by one percent or less, always represents a triumph of good over evil?

Neither company appears hopeful about extending the market for organic foods to the bottom half of the income scale. *Business Week* says Wal-Mart's Lee Scott is "determined to get affluent customers to spend more [on products like organic food] when they come in to buy basics like detergent at Wal-Mart."[27] Meanwhile, Mackey puts his faith in economic progress: "I think [organic and natural food] can continue to penetrate, as the culture becomes wealthier."[28] Whether or not the culture becomes wealthier, the traditionally well-heeled Whole Foods customer base probably will. But if people inhabiting the culture's less elegant precincts are to become wealthier as well, they're unlikely to do it by growing food for big-box food retailers like Wal-Mart or Whole Foods or by taking jobs there.

OTHER ROUTES

In his book, Pollan traced food webs of varying righteousness, the ideal represented by a meal composed almost completely of foods supplied by Pollan himself. But, as he acknowledged, everyone in what we think of as the real world has to depend on lots of other people and resources for daily food. The cashiers in my Wal-Mart and Whole Foods scenarios would certainly have benefited from having their own vegetable gardens. But unless, against all odds, they also managed to raise a lot of staple foods on their own—wheat, oats, dry beans, maybe some chickens or dairy animals—or had plenty of time for fishing, they would still be largely reliant on purchased food of some kind.

Academics and grassroots activists in the sustainable-agriculture movement are doing some hard thinking about how society can pay farmers (preferably small-scale farmers) adequately to raise nutritious food in less ecologically destructive ways while greatly shortening food-transportation paths. A secondary goal has been to keep truly whole food affordable for all, but the ability of the food-producer to earn a good living has always come first.

Like many in her profession, Rhonda Janke, associate professor of horticulture and a sustainable cropping systems specialist at Kansas State University, is a big advocate of locally produced food, farmers' markets and community supported agriculture, or CSA. (In a typical CSA arrangement, consumers contract with a farmer in their area to deliver a certain quantity of food on an agreed schedule during the growing season. The kinds of foods delivered depends on what's in season.) As for making good, locally produced food affordable, Janke notes that

> many CSAs have provisions for "work shares" and reduced cost shares for low-income families, and that can be part of a local safety net. But that doesn't eliminate the need for the grower to get full price from at least a minimum number of full-paying customers. What do you consider a living wage for a farmer, and does that make the price of food go up? If all farmers got $10 per hour, not counting federal subsidies, what would food cost? Would the price at Wal-Mart go up? At Whole Foods?[29]

Janke believes that leaving it up to the big retailers would put food out of reach for a lot of people: "I think the only conclusion one can logically come to is that market forces alone are not going to provide enough healthy food to everyone in our society."

In that spirit, a growing number of non-profit organizations have launched pioneering efforts to put good food within reach of their local communities. One of them is People's Grocery in Oakland, Calif. The community-based organization sells fresh produce and staples through its store and Mobile Market—a "grocery store on wheels" that travels through West Oakland making regular stops. The organization also has extensive educational programs and has helped establish a growing network of community gardens that currently provide 25 percent of the produce it sells.

I asked co-founder Brahm Ahmadi what makes it possible for People's Grocery to sell good, natural food that low-income families can afford, while Whole Foods can't. He said the fundamental difference is that "they're pursuing profit and we aren't. They are purely profit-driven, so they do not allow that cost benefit to go to the customer." Once, says Ahmadi, a Whole Foods executive told him: "We could not market food the way you do, because our shareholders simply wouldn't allow it."[30]

The growth of the natural-food industry has been extraordinary, but Ahmadi predicts that its relatively affluent target market cannot avoid saturation. While Wal-Mart gears up to penetrate the non-affluent market, People's Grocery is already there, subsidizing its efforts through charitable funding, with the understanding that the donated money will go to hold prices down. But, says Ahmadi, People's Grocery will try to reduce its dependency on contributions by selling food that it obtains directly from producers, cutting out as many steps of the expensive supply chain as possible.

Indirectly echoing Rhonda Janke's conclusion that "market forces alone are not enough," Ahmadi said, "We need to build demand that can thrive and grow on its own," but in low-income areas "it has to be done differently. It requires a grassroots approach with community organizations that have track records." And community organizations working in downscale urban zip codes from coast to coast are establishing just such track records. They include, among others, Garden-Raised Bounty in Olympia, Wash., Growing Power in Milwaukee and Added Value in Brooklyn.

A GAPING HOLE

No efforts to date, whether big-corporate or community-based, have found a way to fill the biggest hole in the organic market. If you look at the quantity of land required to feed one human—the average citizen of either America or the world—70 percent or more of that land goes to produce cereals like rice, wheat, and corn, food or feed legumes like chickpeas and soybeans, or oilseeds like sunflower and, again, soybeans. (A big share of that harvest is passed through animals in the US, a smaller share in poorer nations.) But the economic underpinnings of organic and natural food, whether they're found in CSAs, farmers' markets, neighborhood groceries, Whole Foods, or Wal-Mart, are fruits and vegetables, each of which occupies only 4 percent of the land area that supports us.[31]

The bulk of Americans' calories and nutrients comes out of vast, sparsely populated regions hundreds or thousands of miles from the coastal bastions of organic dining. As we've seen, it's those rural regions that are being hit with the biggest share of agriculture's ecological fallout. They're also reaping few of its benefits. Of the 500 poorest counties in the United States, more than 450 are rural. With much of the land under extensive monocultures of wheat, corn, soybean, and other grains, or giant feedlots, and in many areas, long, cold winters, the concept of "eating local" is often greeted with grim chuckles in much of middle America. Ironically, it is in those ecologically threatened agricultural regions that supply the lion's share of the nation's food where sustainably produced, nutritious food is least widely available.

There is plenty of money to be made in natural meat, egg, and dairy products, which could have an indirect impact in grain-cropping country. But the natural/organic giants are under fire for their animal-production practices as well. Detractors charge that many of them allow their cattle, hogs, and poultry only the minimum amount of space and free movement required by USDA's organic standards, and that except for the absence of superfluous drugs and chemicals, organic corporate animal husbandry looks an awful lot like any other corporate animal husbandry. In his web exchange with Mackey, Pollan took Whole Foods to task for not insisting that the beef it buys be primarily grass-fed.[32]

Organic farming, which is what most people think of nowadays when they think of more natural alternatives, addresses one aspect of agriculture's destructiveness: chemical pollution. But since the origin of agriculture ten millennia ago, the most devastating plague spread by farming has been soil erosion and general soil and water degradation caused by tillage. The advent of so-called no-till grain farming has slowed soil erosion and has made a small contribution to slowing global warming in the expanding areas where it has been employed (by reducing tractor use and storing up soil organic matter.) Trouble is, without chemical weed-killers, no-till is extremely difficult and usually impossible. In ruling out herbicides, organic farming not only cannot go no-till on a large scale; it is even more dependent on tillage than is conventional agriculture and requires more fuel for more trips through the field than does no-till, especially on large scales and when growing grain crops.[33] Attempts to develop organic no-till methods depend heavily on two additional inputs: ideal weather and plenty of good luck, both of which are in chronically short supply on the farm.

The road to resolving this dilemma of grain farming—chemical no-till versus chemical-free "more-till"—has so far run into a dead-end because of humanity's dependence on shallow-rooted, short-lived annual grain crops. Deep-rooted perennial grasses and legumes like those on which many pasture- and range-fed cattle graze do a beautiful job of keeping soil ecosystems healthy but don't produce harvests of grain that are edible by humans. Long-term breeding programs to develop perennial grains are underway (and I am involved in those efforts),[34] but for the near future, grain farming remains on the horns of the organic-versus-no-till dilemma.

Grains carry a lot of food in a compact, low-moisture, durable, portable package. For those reasons and because they can be harvested in huge quantities, the production and sale of grains, organic or conventional, doesn't generate anything like the profits that can be gotten from organic strawberries or wine. Neither Whole Foods nor Wal-Mart could make a profit if they gave the majority of floor and shelf space to grains and legumes, in proportion to their importance in our diet. It's high-input, high-value, high-maintenance foods that draw customers to the local farmers' market and claim the high profits for big-box organic sellers. It's no surprise that, according to *Forbes*, Whole Foods realizes a significantly higher proportion of its sales in perishables and prepared foods than do regular grocery stores.[35]

Whole Foods sells bread, too, of course. But cereal grains and legumes—readily available, inexpensive commodities without the sensory impact of Shiitake mushrooms or Pur Brebis P'tit Basque cheese—are an unlikely foundation on which to build any eco-capitalist empire. In writing an article on perennial grains for a scientific journal a few years ago,[36] I learned that Indian ricegrass, a perennial grass native to western North America, was being grown profitably as a grain crop in Montana. Its edible seed, it turned out, was finding a place in the gluten-free market. Unable to find much in print about it, I telephoned a person involved in the farmers' cooperative that was growing and marketing ricegrass grain. He was reluctant to provide any production figures and in fact pleaded with me not even to mention Indian ricegrass in the article I was writing. If more people or groups read about the highly profitable project and went into production themselves, he told me, it would drive down prices and ruin the market for everyone. I wrote about ricegrass anyway, and probably as a sign of the small number of people who read my article, ricegrass flour continues to be marketed successfully.[37]

The ricegrass farmers' dilemma exemplifies the eternal problem of commodification in agriculture; to help organic and natural firms to conquer the market, to expand dramatically while staying true to their original principles, the food really would need supernatural powers. Wal-Mart or no Wal-Mart, the big price premiums that created the organic industry will shrink, and it will become harder and harder to make a profit while fully paying all the ecological bills.

Among John Mackey's many oft-repeated comments is this one: "Business is simple. Management's job is to take care of employees. The employees' job is to take care of the customers.

Happy customers take care of the shareholders. It's a virtuous circle."[38] It's easy to keep a circle like that spinning when it's greased by free-flowing cash. But as they expand, and even as they try to reach out to lower-income customers, all corporations in the whole-food business will feel the need to put the squeeze on their own employees and suppliers.

On top of the $175 billion or so that Americans spent for food in 2006, we spent massive sums to equip the kitchens for which it was destined, buying 23 million new cooking ranges, 13 million refrigerators and freezers, and 7.4 million dishwashers.[39] And we spent $3.2 billion to bring 512 million brand new pots and pans into American kitchens.[40] All of the metals and other materials in those products are found as elements or compounds in the earth and pass through factories, retail outlets, and kitchens before returning eventually to the earth. Along the way, some leave their mark on the biosphere.

9

THE WORLD IS YOUR KITCHEN

"Better things for better living—through chemistry." For more than four decades, E.I. DuPont de Nemours and Company lured customers with that slogan, one of the most memorable in American advertising. It was hatched in 1939 by a public relations firm that DuPont had hired to make the company look less like an explosives-maker (a role that brought it wealth and power during the First World War) and more like a developer of convenient consumer products.[1]

The abundance of twentieth-century industrial chemicals pumped out by DuPont and other companies over the years has done much to make modern life easier and, in a way, better. Pharmaceuticals and pesticides have always gotten the lion's share of press, both good and bad, on chemicals. But other, lower-profile compounds that go into making routine household, farm, factory, and office goods truly transformed life in the industrial West and eventually across the globe. Unlike drugs and bug-killers, these compounds have been widely assumed to be biologically inactive. But over the years, society has learned the hard way that chemical-dependent "better living" can sometimes come at a high biological price.

SOME OF THE PLANET'S TOUGHEST LITTLE MOLECULES

In the 1960s and 1970s, chemicals like polychlorinated biphenols (PCBs)—used as lubricants, adhesives, plasticizers, and insulators, most prominently in electrical transformers—made some highly toxic headlines and were eventually regulated and banned. More recently, two groups of DuPont products developed during that same era, fluorotelomers and fluoropolymers, have come under attack. They are not as strongly toxic as PCBs or dioxins, but their still-unfinished story demonstrates once again that the less that industry, governments, and the people know about the troubling effects of highly profitable products, the more smoothly the economy grows.

Fluorotelomers and fluoropolymers go into making an extraordinarily wide range of products, including non-stick cookware, grease-resistant food packaging, stain-resistant fabrics and carpets, shampoos, conditioners, cleaning products, electronic components, paints, firefighting foams, and a host of other artifacts of modern life. But like many "better things" produced by industrial chemistry, these compounds have opportunities throughout their existence, from their birth in a factory through their eventual chemical decomposition, to invade the biosphere. Because the "better things" those compounds are used in making—products like the nonstick pans in your kitchen, the popcorn bag you toss into the microwave, the box the pizza delivery guy hands you, and the stain-resistant carpet that repels the sauce that falls from the pizza—are designed for contact with food and the human body, there's been a lot of concern over their direct impacts on health. But the more serious ecological hazard may be posed by other, related compounds that are used in the manufacture of those commodities or escape into the environment as they decompose. These disreputable relatives of fluorotelomers and fluoropolymers, fellow members of the perfluorochemical (PFC) family, are turning out to be avid and hardy travelers, showing up in the bloodstreams of people and other animals across the globe.

Some of the same chemical properties that make PFCs so useful in industry make them virtually indestructible in nature. For sheer persistence, two members of the family, perfluorooctanoic acid (PFOA) and perfluorooctane sulfonate (PFOS), are real standouts. They are not broken down by heat, light, or microbes.[2] Other PFCs do break down, but in doing so, many of them end up giving off PFOA or PFOS.

Those compounds and/or other PFCs have turned up in wildlife on at least three continents and above the Arctic Circle, in the blood of dolphins, seals, sea lions, minks, otters, polar bears, gulls, albatrosses, bald eagles, sea turtles, penguins, and dozens more species.[3] They are widespread in seafood.[4] Fifteen PFCs have been identified in human blood samples, with highest concentrations in North Americans.[5] PFOA is the most widely studied of the compounds and has been found in the blood of 90 to 95 percent of US residents who have been screened.[6] In a study of tea workers in Sri Lanka—a group of people unlikely to eat a lot of microwave popcorn—all blood and semen samples contained PFOA, PFOS, and other PFCs.[7]

PFCs thus join hundreds of other complex synthetic molecules that have made their way into the living world over the past half-century. The extent of this mass invasion has been only sparsely monitored. Surveys documented in the US government's Centers for Disease Control and Prevention's (CDC's) 2005 *National Report on Human Exposure to Environmental Chemicals* looked for 148 industrial elements or compounds in the blood and urine of US residents and found most of them.[8] And, as PFCs demonstrate, a molecule that infiltrates our own species can't be kept out of others. How communities, populations, and individual organisms will respond to the onslaught of exotic compounds—molecules not encountered during the evolution of today's species—is not clear.

Suppose two or three independent studies show that an individual chemical has little or no statistical association with disease or death in humans or lab animals—does that mean we can move on to enjoy all the benefits it offers, in unlimited helpings? Might a fourth test, maybe with different species or in combination with other chemicals, show toxicity? Can we confidently go on creating and unleashing novel chemical compounds, their innocence assured at best by a few lab studies that are usually funded or conducted by the companies making them? Or is there something inherently wrong with the accumulation of alien chemicals in living things, in ever greater quantities?

The oft-cited but too-seldom followed "precautionary principle"[9] says that a new synthetic chemical, indeed any new technology, should be considered guilty until proven innocent.

According to the principle, no compound should be approved for general use until it has been thoroughly tested and found safe. That is considered a highly conservative approach, but it shouldn't be seen as guaranteed protection. Some of the most important effects of a compound, especially its impact on whole ecosystems, may not be detectable until it's been around the world a few times. The only way to evaluate chemicals fully is to release them wholesale into the ecosphere—and by that point, it's too late to retrieve persistent ones like PFOA and PFOS. As we will see, that's the kind of "testing" that's often done, the kind that keeps business humming along and industry happy. The alternative—to consider the approval of any compound as only tentative, to monitor its effects at every stage, and to pull the approval at the first sign of problems—would dramatically slow the development of new products. To do that, in industry's view, would be to push the economy toward a steep, nonstick slope.

TURNING UP THE HEAT ON TEFLON

Polytetrafluoroethylene, known by the brand name Teflon, is not a complicated molecule at all—a long, long string of carbon atoms each attached to two fluorine atoms. But it's the slipperiest material ever discovered, and as a result, it has seen phenomenal commercial success not only in industrial applications but also, and most prominently, in the kitchen. Fifty-four percent of US cookware is nonstick, much of it coated with Teflon.[10] Slick pots and pans are among the many kitchen technologies that the US Department of Agriculture credits with improving the status of women over the past half-century.[11] Among other virtues, nonstick cooking surfaces make low-fat and even fat-free cooking much easier, and so can be credited with helping keep arteries clearer than they would be if all frying-pans and griddles were kept slick with oil.[12]

However, sporadic reports of pet birds dropping dead[13] and humans developing flu-like symptoms after breathing fumes from overheated fluoropolymers, including Teflon,[14] have raised concerns since the 1950s. In one case, within three days of placing

2,400 broiler chicks in a poultry research facility, researchers found more than half the chicks dead from "pulmonary edema caused by exposure to an unidentified noxious gas." The cause was traced to Teflon-coated heating lamps.[15] Even Teflon-based ski wax has caused problems.[16]

In 2007, *Consumer Reports* magazine set many minds at ease when it reported that Teflon pans do not emit PFOA when heated above 400 degrees Fahrenheit.[17] But that result is exactly what would be expected. It is at temperatures higher than 400 that Teflon has been shown to release organic fluorine compounds, and PFOA has never been found among the ten or so compounds identified. Some of the gases that are released are acutely toxic.[18] A 2001 study found that under very high heat, trifluoroacetate and other greenhouse gases can be emitted and concluded that "continued use of fluoropolymers may also exacerbate stratospheric ozone-depletion and global warming."[19] Normal cooking doesn't heat Teflon to the temperatures used in that study, but the authors note that the "pre-application heating" of Teflon to such levels is common, and because of its extreme heat resistance, the polymer's industrial applications go to higher temperatures.[20]

The government has never issued warnings about the safety of Teflon, and DuPont officials have always disputed Teflon's toxicity, arguing that when used at proper temperatures under normal kitchen conditions, Teflon cookware does not release dangerous fumes. The company's Teflon-safety web page states: "DuPont non-stick coatings will not begin to significantly decompose until temperatures exceed about 660°F (349°C)—well above the smoke point for cooking oil, fats or butter. It is therefore unlikely that decomposition temperatures for non-stick cookware would be reached while cooking, without burning food to an inedible state."[21] A vice-president of the Cookware Manufacturers Association took the issue right to the bottom line, telling one reporter: "There's no evidence of safety concern whatsoever with using nonstick pans, and sales figures prove that."[22]

Whether or not industry officials are justified in holding out for the absolute safety of Teflon cookware, they appear to be correct when they claim that it's not the major direct source of the PFCs that are so widespread in human blood. Emissions of PFCs during manufacture of Teflon and other fluoropolymer products, and their release from fluorotelomer products like food packaging and fabric treatments, may be more important sources. But there is still much uncertainty over why, today, most of us bleed and pee PFCs.

The eight-carbon molecule PFOA is used as a processing aid in the manufacture of Teflon—helping distribute the compound evenly over a surface—but it is not supposed to be part of the final product. DuPont began producing its own PFOA in 2002, when the 3M Company, its supplier until that time, halted production in response to public pressure. Newborn litters of laboratory rats whose mothers were fed PFOA have experienced higher death rates, slower development, lower body weight, and abnormally fast sexual maturation.[23] Mammary glands of both mothers and daughters developed abnormally.[24] These effects occurred despite the fact that female rats excrete PFOA from their bodies much more efficiently than do humans. Swedish scientists have shown that PFOA may have the potential to compromise immune systems[25] and that rats fed a normal diet had no liver cancer while one-third of rats fed PFOA developed malignant liver tumors.[26] Tumors of the liver, pancreas, mammary glands, and testicles have been associated with PFOA.[27] Published studies on rats, led by DuPont and 3M scientists, found only minor reproductive effects[28] and concluded that PFOA may not pose a cancer risk to people because the kinds of tumors that were found in rats develop by a mechanism that doesn't occur in humans.[29]

DuPont devotes an extensive portion of its company website to PFOA and is clear in its conclusion: "To date, there are no human health effects known to be caused by PFOA. Based on health and toxicological studies conducted by DuPont and other researchers, DuPont believes the weight of evidence indicates that PFOA exposure does not pose a health risk to the general public."[30]

PFOA's cousin PFOS has also shown serious effects on lab animals, along with a strong tendency to concentrate in humans and the wildlife food chain.[31] 3M Company phased that chemical out of its big-selling Scotchgard fabric protector and other products in 2000–2. Other PFCs have undergone far less testing, but some are suspected of being toxic.

Studies of PFOA's and PFOS's effects on humans have depended on comparing the health of people—usually company workers—who've been heavily exposed to the chemical with the health of people who've had no more exposure than the average US resident. DuPont and 3M officials have insisted, based on such studies, that no statistically significant relationships between PFOA exposure and disease or death have been seen in humans.[32] But studies of chemical workers completed so far were all done by the companies themselves. The Environmental Working Group (EWG), an industry watchdog in Washington DC, doesn't claim that employee-health studies carried out by 3M and DuPont were intentionally biased. But EWG does maintain that the studies had flaws in their experimental designs that tended to make it harder to show that effects of PFCs were statistically significant.[33] Meanwhile, 3M and DuPont scientists argue that experiments reporting possible associations between PFOA exposure and prostate or bladder cancer in workers are statistically weak.[34]

A 1984 internal DuPont memo showed that officials considered eliminating PFOA emissions at that time. Discovered almost 20 years after it was written, the memo said in part: "Looking ahead, legal and medical will most likely take a position of total elimination. They have no incentive to take any other position. The product group will take a position that the business cannot afford it."[35] DuPont went on to double its PFOA emissions over the next 15 years.[36]

Whether PFCs are getting into our bodies from Teflon pans on burners we've forgotten to turn off or from home-delivered deep-dish pizzas or from water contaminated by a fluoropolymer manufacturing plant, it is their potential to harm human health that has stirred governments and citizens to action. Over the past few years in the US, the heat on DuPont has been gradually turned up from "simmer" to "high," with lawsuits and proposed regulatory actions in states across the nation. The following is a long list, meant to impress you with the sheer number of hazards revealed and struggles undertaken simply to get a single PFC compound, PFOA, under voluntary regulation:

- In August, 2001, DuPont settled a lawsuit that was brought by members of a family living near the company's Washington Works plant near Parkersburg, West Virginia after they saw more than 250 of their cattle die from horrific physical and mental symptoms.[37] The company continued to maintain that its chemicals did not sicken the cattle, and no link was ever proven.

- From 2003 to 2006, mostly small amounts of PFOA were detected going into groundwater and the Cape Fear River near DuPont's Fayetteville Works plant in North Carolina, at that time the company's facility for manufacturing PFOA salts. Local opposition grew when, in 2005, water in a well close to the plant showed a much higher level of 765 parts per billion (ppb).[38]

- The 2004 settlement of a class-action lawsuit brought by Ohio and West Virginia residents living in the vicinity of the Washington Works plant required the company to spend more than $100 million to ensure that homes in the area are supplied with water uncontaminated with PFOA.[39]

- That settlement also created the C-8 Health Project, a 5-year study correlating PFOA blood-serum levels in more than 60,000 area residents with nine types of medical conditions, starting with cancer, heart disease, and birth defects.[40] ('C-8' is another term for PFOA.) In May 2007, scientists performing the study announced that reporting of results would be delayed until at least the end of 2007.[41]

- A court-appointed science panel of three prominent epidemiologists advising the C-8 Project requested permission in the fall of 2006 to study the effects of PFOA on nearly 5,000 Washington Works employees, many of whom have extremely high blood PFOA levels. They were backed by the United Steelworkers' Union, which represents the workers. DuPont fought to keep employees out of the study, claiming that its own worker surveys have demonstrated that there are no health risks.[42]

- EPA sued DuPont in 2004, charging that the company had for years been concealing information on PFOA pollution at Washington Works. In early 2005, without admitting any wrongdoing, DuPont agreed to pay $16.5 million in fines and support of research and education—the largest civil judgment in EPA's history to that point.[43]

- In January 2006, a Scientific Advisory Board appointed by EPA to review the agency's risk assessment of PFOA voted, by a 12–4 majority, to recommend labeling PFOA as "likely to be carcinogenic" in humans, based on animal studies. DuPont disputes the designation, and EPA has not included it in its as-yet unfinished assessment.[44]

- In April 2006, residents of the area around DuPont's Chamber Works plant in Salem Co., New Jersey filed a lawsuit claiming that the plant had contaminated their water supply with PFCs and that the company had known for years that it was doing so. DuPont said the suit was "without merit."[45]

- At about the same time, class-action cases against DuPont in twelve states were consolidated as one big case in federal court in Iowa, alleging that the company did not inform consumers that it knew Teflon can emit harmful fumes when overheated. DuPont denied there was a problem.[46]

- In November 2006, a committee of California's Office of Environmental Health Hazard Assessment met to consider making PFOA a top-priority chemical for review as a possible carcinogen. The committee was divided on the issue, but if the agency decides to designate PFOA as cancer-causing, any product containing the chemical will have to be so labeled. DuPont is fighting to keep the label off its products.[47]

- In a consent order that same month, EPA forced DuPont to agree that if the water supply of any household near Washington Works showed a PFOA concentration above 0.5

ppb, the company would pay to provide water treatment or an alternative water supply.[48] EPA explained the necessity for its 2006 consent order like this:

In a 2002 consent order, DuPont agreed to provide residents with bottled water or install water treatment equipment when the level of C-8 in drinking water was measured at 150 ppb [parts per billion]. Recent studies, however, show that people who live in the vicinity of DuPont's Washington Works plant near Parkersburg, W. Va. have a median C-8 level of 298 to 369 ppb in their bloodstreams. This is much higher than the 5.0 ppb found on average in the blood of the general U.S. population. New studies have demonstrated various kinds of toxic effects on experimental animals, and the results are a concern for public health. Therefore, the new consent order between EPA and DuPont that took effect Nov. 20, 2006 lowers the "action level" to 0.50 ppb.[49]

As we'll see, DuPont has finally set a goal of eliminating the use of PFOA altogether. But without the tenacious efforts of Environmental Working Group in exposing the problem; the United Steelworkers' actions in defense of its members; communities that wouldn't accept the fact that they can no longer drink their own water; a family that lost an entire herd of cattle; panels of principled scientists who refuse to ignore hard data; and other less well-publicized groups, DuPont might be releasing PFCs without hindrance. EPA appears to take pride in the fact that DuPont has been brought partially to heel without direct regulatory action. It has taken people working collectively and spontaneously to push the government to do just one small part of what it should have done in the course of its normal duties.

This is a societal, not just a personal, issue. True, the most prominent products made with fluorotelomers and fluoropolymers are quintessential consumer goods, and it's the accumulation of millions of purchases at supermarkets and big-box stores that creates the demand for PFCs. But like many ecological problems with a domestic face, this one will never be successfully attacked by relying on consumers to go shopping for better health. Dramatically elevated levels of PFCs were found not in home cooking but in the blood of 3M and DuPont workers and people living near their plants. Long-term consequences may eventually be seen anywhere from the ice floes of northern Siberia to the fisheries of the Gulf of Mexico. And the range of products containing or releasing PFCs is so wide and diverse that consumers would be hard-pressed to "vote with their pocketbooks" even if they did manage to identify all items involved (and even if they decided they could live without them.)

CHEMICAL STEWARDSHIP

In early 2006, DuPont and seven other companies signed on to EPA's voluntary "PFOA Stewardship Program," under which they pledged to reduce emissions of PFOA from their factories by 95 percent by the year 2010.[50] Then, in a 5 February 2007 press release, DuPont officials announced that they had reduced the residual PFOA content of their products by 97 percent and they expected to reduce factory emissions by 97 percent before the end of 2007. "In addition," Dupont chairman Charles O. Holliday, Jr. declared, "we are developing potential alternative technologies, and today we are committing to eliminate the need to make, buy or use PFOA by 2015."[51] The day after the announcement, analyst Kristan Markey of EWG told me: "This was a great step forward. We just wish they'd done it years ago." Although "DuPont believes that with the Stewardship Program, they have solved the problem," Markey said "huge open questions" have been left hanging: How did so much of the stuff get into the environment, and what does that imply for the future?[52]

DuPont may be able to walk away from past pollution, but, said Markey, "We have two key contentions with that." Firstly, he noted, "DuPont has conveniently forgotten that vast quantities of PFOA-emitting products [like food packaging] that they and other companies sold over the years have not disappeared." Either already lying in landfills or destined to end up there, they can continue giving off PFOA. Secondly, he asked, "Will fluoropolymers [like Teflon] also break down to release PFCs? Academic and independent research is in progress to answer that question. If I were a betting man, my money would go on a finding that they do break down." So even if PFOA is eliminated from factory emissions and products, the indestructible compound could continue to spread through the environment, through release from fluorotelomer products like popcorn bags and carpets and potential release from the breakdown of fluropolymers.

When a company announces an environmental breakthrough, it often involves the discovery of a new material or process that allows production and sales to continue uninterrupted. That, to cite a prominent example, is what differentiates the problem of ozone depletion (on which progress has been made because substitute refrigerants were developed) from global warming (which economies have failed to curb because no direct substitutes have been found that have the concentrated energy and versatility of fossil fuels). DuPont insisted upon continuing the manufacture and release of PFOA for decades, in the face of accumulating evidence of its potential for harm. Finally, a way was found, as DuPont executive Holliday said in the press release, to eliminate PFOA while ensuring "the continued availability of fluorotelomers and fluoropolymers for essential products used in telecommunications, aerospace, semiconductors, fire fighting and consumer applications." If satisfactory substitutes could not have been found, would DuPont have decided in 2006 to eliminate PFOA in the interest of public health (which they've never acknowledged was imperiled in the first place)? What will be their response if the new products and processes turn out to be troublesome?

It's hard to compel corporations to rein in persistent, widespread pollutants unless it can be shown that they could sicken or kill humans directly. Any effort to oppose economically important substances solely on the basis of broad, complex ecological reasoning is usually doomed from the start (the 1970s ban on DDT in many countries, including the US, is a striking exception, and today's far right still harbors an intense hatred of Rachel Carson, author of the 1964 book *Silent Spring* that helped prompt the ban. Right-wing commentators routinely and wrongly hold her responsible for millions of malaria deaths that they say amount to "genocide" on a scale comparable to those attributed to Hitler, Stalin, and Pol Pot.[53])

A century of research in ecology has taught us that the well-being of humans depends not only on freedom from disease but also on the life-support provided by well-functioning ecosystems. The comprehensive and highly regarded Millennium Ecosystem Assessment emphasized that regardless of the value one might place on species other than our own, humanity itself is in very deep trouble unless we relieve the pressure we're placing on ecosystems worldwide and allow them to restore their ability to function.[54] Human health depends not only on keeping synthetic toxins out of our own bodies, but also on keeping them out of general biological circulation.

Any introduction of alien molecules onto the planet should carry a heavy load of responsibility. But as scientists have observed the general buildup of PFCs in the biosphere, they have learned much less about what that will mean for ecosystems. PFCs aren't acutely toxic; nevertheless, there's little reason for complacency when compounds that can now be found so readily in such a broad range of animals—mammals, birds, and reptiles—have also been shown to damage the health of the few laboratory species on which they have been tested.

Some effects on simple ecosystems have indeed been detected. One 2003 Danish study showed that when PFOA was added to a laboratory ecosystem of freshwater zooplankton (microscopic animals), the system "was changed from a more diverse community dominated by larger species towards a less diverse community dominated by smaller and more robust species."[55] Multiply that tiny experiment involving only one group of species by the billions of communities of plants, animals, and microbes that are now dealing with PFCs in their midst, and it requires a great deal of faith to believe that we aren't knocking the planet at least a little off-balance.

PATHS OF LEAST RESISTANCE

Starting in 2006, once a day on average, a tanker truck loaded with a PFC called fluorotelomer alcohol departed DuPont's Chamber Works plant in New Jersey en route to the company's First Chemical plant near Pascagoula, Mississippi. There, PFOA contaminant was removed from the fluorotelomer alcohol, and the purified product hauled back to New Jersey. DuPont said a small amount of escaped PFOA, 2 pounds per year, would go into the area's sewer system, which ultimately flows into the Gulf of Mexico. Local officials were reportedly "surprised and disappointed" that the company did not inform them that it was beginning the process.[56] PFOA is not a regulated substance (all restrictions slapped on it to date have resulted from civil settlements, not regulatory actions), so emissions from the First Chemical plant require no EPA or state permits. Not to worry; DuPont said that it would do its own monitoring and report the results.[57]

Brenda Songy, a member of the Mississippi Sierra Club, was not reassured. She'd been battling from the beginning to stop the importation of PFOA from New Jersey. As part of that effort, Songy and other opponents attended a meeting of the Pascagoula City Council in October 2006, to urge a ban on emissions of the chemical into city sewers. But, Songy told me a couple of weeks later, city officials refused even to consider turning away the tanker trucks. Meanwhile, neither the city nor anyone else outside DuPont knows what process the company is using to remove PFOA: "They say they can't tell us, that it's proprietary information."[58]

In Songy's opinion, the city and region are "too entrenched in industrialization" to resist the importation. That's despite the huge load of environmental toxins and high cancer rates that past industry-friendliness has brought Pascagoula. She said: "They didn't put it in these words, but the message I got from them was, 'Go on home and take care of your baby. We big boys will take care of this issue.'" All the council members could see, she said, was the $20 million the company was investing to retool First Chemical to handle PFOA removal. Referring to the city's leaders, Songy emphasized, "They aren't bad people. They're kind-hearted, and they really believe these companies are also kind-hearted, that they would never put our lives at risk." But at the time we spoke, she was making plans to pick up and move away from Pascagoula.

First Chemical plant manager James Freeman explained to the South Mississippi *Sun Herald* that DuPont was going to such lengths—hauling loads of fluorotelomer alcohol more than a thousand miles, day after day—because his plant already has much of the equipment needed to scrub PFOA from the alcohol. It would be "impractical and too expensive," he said, to haul the necessary equipment to New Jersey or buy it new.[59] Freeman did not cite other reasons for the company's seemingly cumbersome solution to PFOA contamination. For example, might opposition by Jerseyites, including their pending lawsuit against DuPont, have convinced the company not to decontaminate the fluorotelomer alcohol right there at its site of origin?

Pascagoula's investment-friendly atmosphere, which placed it right in the path of least resistance, may result in part from the area's economic weakness, as compared with Salem Co., New Jersey. Table 9.1 compares poverty rates and median incomes of the counties affected by the DuPont facilities mentioned in this

Table 9.1 Poverty and income levels near some DuPont facilities involved with fluorotelomers and fluoropolymers.[60] (Washington Works is located next to a state line and potentially affects two counties.)

DuPont facility	Environs	Percent below poverty line, 2004	Median household income, 2004
Washington Works	Wood Co., West Virginia	20.4	$36,462
Washington Works	Washington Co., Ohio	18.2	$36,297
Fayetteville Works	Cumberland Co., North Carolina	21.5	$39,035
Chamber Works	Salem Co., New Jersey	9.2	$49,231
First Chemical	Jackson Co., Mississippi	23.1	$40,418
Corporate Headquarters	New Castle Co., Delaware	8.9	$54,304

chapter with those prevailing near its corporate headquarters. (For comparison, in 2004 for the United States as a whole, 12 percent of the population lived below the poverty line, and the median income was $43,389.)

Corporations have long been known to locate their more controversial activities in more economically hungry places. For decades, the environmental justice movement has been drawing attention to the fact that heavily polluting industries or companies' dirtiest factories are most often sited in communities where economic need is great enough to trump concerns over health and the environment. One of many studies backing up that claim is a 2003 report by the Oakland-based Environmental Justice and Health Union, which found, based on CDC's surveys of bodily contaminants, that non-Latino black Americans are much more heavily exposed to dioxins and PCBs, and synthetic chemicals in general, than are other ethnic groups.[61] But if companies do indeed intentionally target poorer localities, it doesn't guarantee that they'll always get their way. Despite economic distress, all of the affected communities listed in Table 9.1 have mounted resistance of some kind.

THE CHEMICAL AMNESTY PROGRAM

Governments, including the US government, have only a few legal means by which they can put limits on the products turned out by industry. Their hand is strengthened when science shows that a product causes illness, injury, or death to people or other mammals. But simply demonstrating a health hazard isn't enough. Governments are under a strict obligation to balance their duty to protect citizens and the environment with their duty to maintain strong economic growth. Even when they're holding an ace— research results predicting increased sickness or mortality, with cancer as the ace of spades—government agencies generally attempt regulation only if they can show that a product's potential to damage human health significantly exceeds the potential costs to business of having it regulated.

Such cost-benefit analyses require, logically, that a dollar value be placed on each human life that's expected to be saved or lengthened. That quickly leads into some dark territory; as economist and life-valuing expert Ike Brannon has noted: "It is not uncommon for well-meaning people to object strenuously to placing a value on a human life, judging such a practice to be callous and demeaning the value of existence." On the other hand, says Brannon, people make such unsavory decisions in an informal way all the time: "... we do not all drive armored cars to work, but instead drive somewhat less safe—and considerably less expensive—cars."[62]

With actual markets in human beings (officially) a thing of the past, calculating the price of a life has been left up to governments, insurance companies, and other interested parties. While most people would be reluctant to place a dollar value on a life, let alone, say, one additional year of life at, say, 75 percent of full health, big business shows no reluctance at all to express its views on the subject. At industry's urging, the George W. Bush administration put relentless pressure on EPA not only to use methods that give low dollar figures—in one case, slashing the value of an average life from $6 million to $3.7 million—but also put a discount on lives of older people.[63] Doing so serves business interests by making proposed regulations look more expensive relative to the benefits they provide.

EPA has not completed its hazard assessment of PFOA and has done even less work on other PFCs, so it has not even had the opportunity to do cost-benefit analyses for those compounds. If and when they are done, the results will favor industry—you can bet your fluorotelomer-treated trail-running shoes on that. Any estimated dollar value of lives to be lost will probably be small when compared with the value of profits and jobs that chemical producers and chemical-using companies stand to lose if the EPA bans PFCs. Keeping that single class of compounds out of our blood would affect a myriad of products that either are made with or emit them. That's one big advantage that comes from industry's freedom to market chemical products without advance testing: By the time their safety is questioned, the products may have become too economically important to be restricted.

Unlike pharmaceuticals and agricultural chemicals, industrial chemicals receive only minimal federal scrutiny before being OKed for use. Upon passage of the Toxic Substances Control Act (TSCA) in the 1970s, more than 63,000 such chemicals—including PFOA and other PFCs—were "grandfathered in," receiving unrestricted approval for use in industrial processes.[64] Today, between 80,000 and 100,000 industrial chemicals may be used freely in this country; since passage of TSCA, only a handful of such chemical compounds have been placed under restrictions.[65]

A 2006 series in the Fort Worth *Star-Telegram* showed just how loose federal regulation of industrial chemicals is.[66] Under TSCA, according to the paper, the EPA receives an average of 142 new-chemical applications each month, and its too-small staff has only 90 days to review any scientific data bearing on a new chemical's toxicity or persistence in the environment. There is little or no such information in most cases anyway, and neither the agency nor the applicant is required to do any testing before a chemical is approved. Companies are required to inform the government if they know of any adverse data, but before any chemical can be rejected, the burden is placed squarely on EPA to prove that it poses an "unreasonable risk to health or the environment." How high must a risk be to be considered "unreasonable"? That question is left unanswered.

To subject industrial chemicals to the kind of extensive and expensive testing that drugs and pesticides are required to undergo would surely cripple industry's ability to turn out new products as fast as they do. But a system much stricter than the current non-system, one that required environmental studies both before and after a compound is approved, would force companies to be much more deliberate, and maybe even to ask, "Does humanity really need this product?" before turning yet another new synthetic molecule loose on the biosphere.

Industry might tolerate such restrictions now and then but would most likely declare itself totally crippled if society insisted on rigorous environmental evaluation of every synthetic chemical old and new, whatever its nature or purpose. Pharmaceutical companies never stop whining about extensive testing requirements, blaming them for the limited number and high price of products they release each year. Were regulators to take a highly conservative attitude, as they should, toward granting approval not only for drugs but also for run-of-the-mill industrial compounds, the corporate wailing in Washington would be heard all the way to Pascagoula.

In June 2005, the US Government Accountability Office complained about the innocent-until-proven-guilty attitude that the government takes toward new industrial chemicals,[67] but no changes appear imminent. In fact, despite years of controversy over health issues and concern about the widespread presence of PFCs in blood samples, some PFCs are now under consideration as human-blood substitutes to be used in transfusions. No kidding.[68] Meanwhile, environmental and health testing will be conducted as it's been done throughout the age of "better living through chemistry": by exposing humans, other animals, plants, and microbes to molecules that have never before existed on Earth and investigating the consequences only when they're too dramatic to ignore.

In a product-testing regime like that, the most successful compounds, the ones that are produced and consumed in the biggest volumes and varieties of products, get the most thorough testing. But before the results are even in, some of them—like PFCs—can become widespread and near-permanent residents of the planet. So as matters stand today, when corporate executives make their regular announcements that this or that chemical is completely harmless, we can only hope that they're right.

In the kinds of official cost-benefit analyses discussed above, some number, X, of human lives valued at $X million or $3X million or $6X million is set against a cost of $Y million that companies would have to pay to prevent the harm. But, in the case of PFCs, if we were guided by realism rather than economics, we'd drop the dollar signs and the single-minded fixation on human medical conditions, identify the cost as the permanent and ever-increasing release into the ecosphere of alien molecules with demonstrated biological activity, and then set against that a list of what we've gained: easy cleaning of pots and pans, quick popcorn, grease that stays in the pizza instead of on the box, comfortable outdoor gear, slightly better or cheaper computer chips, shinier, more manageable hair, and of course, sustained profitability for big corporations. For the majority of citizens, that analysis would give a very different result.

We now come to the question of what lies behind the economic system's tolerance of alien industrial chemicals, its treatment of natural food as a luxury, its lust for outdated, fossil-fuel-based development, its rationing of vital products like natural gas based on the ability to pay, its parallel abuses of people and the land, its tendency to "solve" health problems caused by consumption with more consumption, its hard-selling of drugs to people at one end of the socioeconomic scale that results in sickness at the other, and its nonstop malignant growth.

10

POLITICAL IMPOSSIBILITY vs.
BIOLOGICAL IMPOSSIBILITY

A book entitled *Green to Gold: How Smart Companies Use Environmental Strategy to Innovate, Create Value, and Build Competitive Advantage*,[1] by Daniel Esty and Andrew Winston, caught the imagination of enlightened capitalists from coast to coast when it was published in 2006. It was a how-to manual for business people wanting to emulate the best examples of "green" capitalism available: companies that, in the authors' phrase, "get ahead of the Green Wave," whose "environmental strategies provide added degrees of freedom to operate, profit, and grow." They quote with approval Wal-Mart CEO Lee Scott, who told his fellow executives "that their sustainability efforts would help protect the company's 'license to grow'." These are some of the helpful tips to be found in this guide for the green capitalist:

- "Most successful green marketing starts with the traditional selling points—price, quality, or performance—and only then mentions environmental attributes. Almost always, green should not be the first button to push."

- "Eco-labels can provide legitimate environmental information to a demanding public. But they can also be used as a trade barrier, disadvantaging competitors in the marketplace."

- "Corporate strategy 101 tells us that a company can drive revenues by increasing price or volume. With green products, volume is a much safer route."

- "Partnering gives a company a strong defense against NGO [nongovernmental (non-profit) organization] attacks, but a large part of that defense is the demonstration of genuine progress. We call it **brand inoculation** ..." [their bold]

Right in the first chapter, Esty and Wilson provide US and worldwide rankings of companies they've designated as green "WaveRiders." Number 1 in the international ranking is petroleum giant BP PLC. Their account of how BP arrived at that enviable spot is little more than a description of a masterful public-relations campaign. Today, for example, "despite being in a business with large environmental impacts, the company is now seen as green." The authors conclude: "Here's the real proof. BP's brand value, as measured by experts in measuring intangibles, has jumped significantly." But BP's primary mission is still to earn a profit by selling fossil fuels, so it was not a big surprise when the *Independent* reported in 2005 that the company had been lobbying against substantive proposals then before the US Congress for capping carbon dioxide emissions, instead supporting a move that would have "companies only to try to cut emissions with the promise of tax breaks." [2]

Then, the year after *Green to Gold* was published, EPA exempted BP from what the company regarded as a too-restrictive environmental law, allowing BP to discharge increased quantities of ammonia and other pollutants into Lake Michigan and to continue dumping mercury into the lake. This reportedly was done so that BP could refine heavier Canadian crude oil. [3]

In a chapter section headed, "Perfect is the Enemy of the Good" (that favorite motto of the meek), Esty and Wilson contrast what they see as an exemplary decision by McDonald's—to give its McNuggets a package that was not environmentally offensive enough to drive away eco-conscious customers yet not so "flimsy" that it would annoy conventional customers—with what they see as the too-radical approach of The Body Shop International PLC, whose pursuit of its "environmental and social mission" was "inattentive to economic realities."

Casting the The Body Shop, a UK-based firm specializing in skin-and-hair-care products, in the role of seeker of "the perfect" would probably strike the company's critics as odd, to say the least. Over its history of more than 30 years, the company has carefully cultivated its image as a pioneer of high business ethics but has been dismissed by detractors as being no more than a world leader in faux-green consumerism. John Entine's devastating expose of The Body Shop finally saw print in 2003 after going unpublished for a decade (because the editors of *Vanity Fair* magazine, for whom it was originally written, feared highly restrictive British libel laws).[4] Others have blasted the company's much-publicized relationships with indigenous peoples.[5] The Body Shop was purchased in 2006—most likely after *Green to Gold* had gone to press—by L'Oreal, the world's largest cosmetics company. To date, L'Oreal has refused to sign a proposed international Compact for Safe Cosmetics. The buyout prompted *Ethical Consumer* magazine to drop The Body Shop from a rating of 11 on the magazine's 1-to-20 "Ethiscore" scale all the way down to 2.5.[6]

THREE BIG BOOKS

The preceding chapters have illustrated how the most well-intended businesses—like those whose aim is to put wholesome food on the table, produce lifesaving drugs, keep hospitals sterile, make farmland more fertile, and keep people fit—can undermine those very efforts when their one mandatory obligation is to turn a profit for owners and shareholders. In this book, I have not highlighted the big-ticket consumption and thoughtless waste one sees every day in the shopping malls of the West and the richer districts of the global South. Rather, I have tried to illustrate how the system destroys even as it provides the everyday necessities of human life.

A useful way to view that scheme is through the writings of three economists who, could they be brought back to life for a panel discussion, would doubtless find themselves at loggerheads on many questions but would probably come to an accord on the current state of the planet. Three books, one by each of those thinkers, have proven far more relevant to the planet's future than the tepid *Green to Gold* will ever be: Karl Marx's *Capital*,[7]

Nicholas Georgescu-Roegen's *The Entropy Law and the Economic Process*,[8] and William Stanley Jevons's *The Coal Question*.[9]

In the opening chapters of his magnum opus *Capital*, Marx described the growth cycle at the heart of capitalism, which depends on everyday things, tangible and not, becoming commodities when their "exchange values" are separated from their inherent "use values." A commodity's exchange value—what it's worth in terms of money or other commodities—emerges in the marketplace, and economies handle that process with ease. On the other hand, there's no easy or systematic way to determine use value. To you and me, the inherent value of a bottle of pills, a pound of tofu, a frying pan, or a hard day's work is obvious, even if we can't quantify it. But inherent value doesn't count for much in capitalist economics, where every scrap of stuff is, first and foremost, an exchangeable commodity.

Using the term C to denote a commodity and M for money, Marx contrasted the simple exchange of commodities, or "selling in order to buy," which he symbolized as C–M–C with "buying in order to sell," or M–C–M, which is the function of the capitalist. The first, C–M–C, is simply the way in which people or groups obtain specific things they need; the initial C might be a truckload of wheat a farmer sells to the local grain dealer or the labor power a worker sells to an employer in order to get money to buy necessities, which is symbolized by the second C. But all money is just money; therefore, as Marx noted, M–C–M (that is, buying in order to sell) "at first sight appears purposeless." But, he wrote, the purpose is revealed when the cycle is written M–C–M', where M' is a larger quantity than M. The difference is the additional value generated by labor and legally snatched away by the employer. That gives the process a purpose: the accumulation of capital.[10]

And the process is a circuit. In Marx's words, "Money ends the movement only to begin it again." He argued that "The expansion of value, which is the objective basis or main-spring of the circulation M–C–M, becomes [the capitalist's] subjective aim, and it is only in so far as the appropriation of ever more wealth in the abstract becomes the sole motive of his operations, that he functions as a capitalist."[11] In other words, what we really have is M–C–M'–C–M''–C–M'''..., a cycle that, from capital's point of view, must run as hard and fast as possible and meet no endpoint; it must grow, perpetually.

One often hears it argued that the solution is to curb the excesses of large corporations and live the purportedly simple economic life described in Adam Smith's writings.[12] But capitalism's destructive growth gave rise to rapacious corporations, not the other way around. DuPont and Wal-Mart are not the roots of capitalism's destructiveness; they're just long branches. Well before the era of corporate rule, Marx saw that "One capitalist always kills many."[13] He saw and foresaw the inevitability of the system's growth as well as its concentration:

> The cheapness of commodities depends ... on the productiveness of labor, and this again on the scale of production. Therefore, the larger capitals beat the smaller. ... The smaller capitals, therefore, crowd into spheres of production which Modern Industry has only sporadically or incompletely got hold of. Here competition rages in direct proportion to the number, and in inverse proportion to the magnitudes, of the antagonistic capitals. It always ends in the ruin of many small capitalists whose capitals partly pass into the hands of their conquerors, partly vanish.[14]

It's true that any individual business is free to stick to the slow lane, keep doing what it's doing, not grow, and try to survive. But that just leaves more space in the passing lane.

About a century after the publication of *Capital*, Nicholas Georgescu-Roegen added another layer to the analysis of growth. An economist at Vanderbilt University from 1950 to 1976, Georgescu-Roegen relied as much on physics and biology as on classical economic theories. He argued that in human societies, the entire economic process can be represented by just two factors: a front-door entrance for resources—concentrated energy and "highly ordered" materials—and a back-door exit for disordered, useless (or less useful) wastes. When all's said and done, he argued, an economy's only product is nonmaterial "enjoyment of life," which can be accumulated only as memories. The Second Law of Thermodynamics—the Entropy Law—tells us that the universe as a whole is winding down, inevitably, toward an eventual high-entropy "heat death." And even though in our own little corner of the universe, argued Georgescu-Roegen, human actions can build highly-ordered pockets of low entropy, the sum total of all our economic activities can only accelerate thermodynamic decay. Economies run in only one direction: downhill, with increasingly worn brakes. He wrote:

> The conclusion is straightforward. If we stampede over details, we can say that every baby born now means one human life less in the future. But also every Cadillac produced at any time means fewer lives in the future. Up to this day, the price of technological progress has meant a shift from the more abundant source of low entropy—the solar radiation—to the less abundant one—the earth's mineral resources ... Population pressure and technological progress bring *ceteris paribus* the career of the human species nearer to its end only because both factors cause a speedier decumulation of its dowry. For we must not doubt that, man's [sic] nature being what it is, the destiny of the human species is to choose a truly great but brief, not a long and dull, career.[15]

Marx had also recognized the indispensability of non-human resources; however, he also saw that in capitalist economies, resources in their native state—undisturbed soil, coal lying under the ground—have no value and therefore no place in the economic equations with which he described capitalism. In capitalist economics, it is the efforts of the farmer, the miner, or other working humans that give economic value to what are considered "free gifts of Nature." Georgescu-Roegen added the input of Nature to Marx's focus on the value of labor. By including the value of those inputs in his analysis, he emphasized that whenever they enter the economic process, they are inevitably and irreversibly decayed—that, to repeat, the only outputs are degraded wastes and intangible enjoyment of life. If you take Marx's hard-driving M C-M' cycle, repeat it without pause, and link it to Georgescu-Roegen's concept of thermodynamic decay, the implications are profound and unavoidable: Provided our species survives, there lies somewhere in its future another Stone Age, and the faster our economic growth, the steeper the decline will be. The next Stone Age will be more resource-poor and probably more toxic than the last, and there will be no shot at a comeback.

Once *The Entropy Law and the Economic Process* was widely hailed as a seminal work, most economists stuffed it into their bookshelves to be consulted no more. Its message was simply too dismal even for the "dismal science."[16] But a few, known as ecological economists, have taken Georgescu-Roegen seriously. Starting with Herman Daly—formerly a World Bank economist, today a professor at the University of Maryland—they have spent decades drawing up blueprints for an economic system that could push that second Stone Age into the immeasurably remote future, while ensuring that the long journey is humane, even comfortable.

Daly's bottom line, as expressed in his 1977 classic *Steady State Economics* (of which an expanded second edition was published in 1991) is this: If we're to live within our material means, planet-wide, we must create institutions with the power to (1) limit our species' rate of reproduction to halt or reverse population growth, (2) hold our total resources-to-waste "throughput" down to a strictly sustainable level, and (3) set upper and lower limits on every individual's or household's wealth and income.[17] These policies make up a package; following only one or two of them won't do the job, according to Daly.

Most ecological economists, while heretical within their discipline, are not explicitly anti-capitalist. In a 2004 ecological economics textbook written with Joshua Farley, Daly does not urge the demolition and rebuilding of whole economies, but rather the "stretching and bending" of existing institutions.[18] Georgescu-Roegen, a Romanian-American immigrant, was certainly no socialist, arguing as he did that humanity could never be rid of its elite strata. While recognizing that inequality breeds insupportable growth, most ecological economists reject direct expropriation of wealth and property from those who have the most, preferring instead to put a limit on the human economy's overall physical "throughput" and have the capitalist class pay the costs of its resource use and ecological destruction. The proceeds, it's argued, would provide for the absolute wants of the majority.

But is capitalism the kind of creature that can survive in captivity? The small, powerful class of people who today reap its economic benefits can be counted upon to rush headlong into ecological catastrophe rather than to permit the creation of institutions like those proposed by Daly and Farley. Low-wage industries simply could not accept policies that would limit the human population—policies that would give workers and consumers greater bargaining power. Manufacturers would simply refuse to slash their use of resources, production of goods, and discard of wastes. And, most crucially, the investing class would never agree to limit its accumulation of wealth in favor of the world's impoverished majority.

Trying to hold a capitalist economy together without growth (and the lopsided accumulation of the riches that growth produces) would be like trying to hold the solar system together without gravity. What would an economy be if it used just enough resources and labor power to supply everyone with essential needs plus as much "enjoyment of life" as the planet could handle, but no more? Whatever it might be, it wouldn't be capitalism. Where would you be working if, when you had produced your fair daily contribution to those needs and enjoyment, your workday would be finished? Wherever you were, you wouldn't be in a capitalist enterprise.

As Herman Daly has pointed out, growth provides an easy way out for societies that don't want to face the inevitable conflict between basic human needs at the bottom of the wealth scale and infinite human wants at the top. If you're a politician or pundit, you can't very well tell working people, "I'm afraid that our owning class is going to be wanting bigger income on its investments for a while—well, actually forever—and it's going to have to come out of your paychecks." Instead, you talk about an always-expanding economic pie with bigger slices for everyone.

EFFICIENCY

It's not hard to find technological optimists who argue, appealingly, that we can have continued economic growth to the benefit of all by achieving greater efficiency. More earth-friendly energy sources, materials, and processes, they say, will let businesses generate more monetary wealth while using and abusing less of the material world. Meanwhile, conversion from a manufacturing-based economy to a "service economy" is expected to help generate more "enjoyment of life" without using up and throwing off so much stuff.

If there's one thing all CEOs and environmentalists can agree upon, it's that efficiency is good. But capitalism has a way of turning good things inside out. If a business owner can produce breaded chicken strips or fever thermometers with less expense on energy, materials, labor, or waste disposal, that's efficiency, and it's money in the bank. But it's the job of a good capitalist to get that money out of the bank, ASAP, and invest it in the real world, where it can be used to turn more labor and materials into more money. (No matter if demand is down—there's always advertising.) Using efficiency to make growth less destructive is sort of like playing "whack-a-mole" at the county fair. Knock capital out of circulation here, and it will pop up over there.

Working in the same era as Marx, the more conventional British economist William Stanley Jevons looked at the substance that underpinned his country's then-thriving Industrial Revolution—coal—and worried what would happen when the nation's mines were exhausted. Through the 14 decades since his book *The Coal Question* was published, Jevons has been derided for not foreseeing that an even more powerful economic catalyst—petroleum—would emerge long before Britain's or the world's coal supplies could run out.

But the much more important part of Jevons's analysis was his explanation of *why* the consumption of coal was accelerating. Based on extensive analysis of data from the British economy, he saw that, ironically, the ever-more efficient use of coal was leading to ever-greater consumption, by stimulating economic growth:

In fact, there is hardly a single use of fuel in which a little care, ingenuity, or expenditure of capital may not make a considerable saving. But no one must suppose that the coal thus saved is spared—it is only saved from one use to be employed in others, and the profits gained soon lead to extended employment in many new forms. The several branches of industry are closely interdependent, and progress of any one leads to the progress of nearly all.[19]

Having provided numerous examples involving the ever-improving steam engine, canals, railways, iron-making, and the steel revolution still looming in the future, Jevons turned to a counter-example: the first steam engine, invented by Thomas Savery. It was highly inefficient and, because of its high cost, was never adopted by industry. Jevons noted with a wink that "it consumed no coal, because its rate of consumption was too high."[20]

Jevons's bottom line: "It is wholly a confusion of ideas to suppose that the economical use of fuel is equivalent to a diminished consumption. The very contrary is the truth."[21] And the stimulatory effect of money-saving efficiency continued to propel the world economy long after Old King Coal was dethroned. In this century, Herman Daly has argued forcefully that "'Efficiency first' sounds good, especially when referred to as 'win-win' strategies or more picturesquely as 'picking the low-hanging fruit'. But the problem of 'efficiency first' is what comes second."[22] That, as Jevons showed, is a more revved up, hungrier economy. Daly offers an alternative: "A policy of 'frugality first', however, induces efficiency as a secondary consequence; 'efficiency first' does not induce frugality—it makes frugality less necessary...."[23] Today's efficiency-minded capitalists, whether their money is in coal-fired power plants or organic lettuce, are still compelled either to deny or ignore the analyses of Jevons and Daly.[24]

In trying to show that constantly improving technology gets us more bang for the buck, or, more aptly, more bucks for the Btu, efficiency enthusiasts like to use ratios. A commonly used ratio is the amount of economic activity per unit of a particular resource used—for example, an average person's share of the GDP divided by an average person's energy consumption. Looked at that way, we're making great progress on energy efficiency; Americans got $97 worth of GDP out of each million Btus of energy burned in 2000, compared with only $68 in 1990. That's a reduction of almost 4 percent per year in the consumption of a single type of resource, per dollar generated. But all of the improvement in that ratio is due to a rising GDP, which, being made of money, is infinitely elastic. Per-person energy use (energy being an all-too-real quantity of which the average American already used 30 times as much as the average citizen of India) actually *increased* by 4 percent over that decade. With our rising population, total US energy use grew by 17 percent. Therefore, what's defined as higher efficiency is actually just an imaginary quantity, GDP, that managed to expand faster than a real, limited quantity, the amount of energy that was burned up. So no progress was made in curbing consumption—unless, like some technology optimists, you reason that efficiency at least prevented the planet from going downhill even faster.

In Chapter 1, we saw the depth of frugality—reaching 80 percent or more of current Western consumption—that will be required if projections of global temperature changes and ecosystem loss are anywhere close to correct. That is a far deeper cut than can be achieved through eco-conscious purchasing or improved production practices by enlightened capitalists. Suppose that the goal is to maintain living standards but use efficiency to reduce resource consumption and waste generation by 4 percent every single year for 40 years (which would accumulate to give us our desired 80 percent reduction before 2050). Making an entire economy operate 4 percent more efficiently—to be truly more efficient, not just greenwashed—every year for four decades would be hard enough. But that's a stagnant, no-growth economy. Allow for modest real growth in the GDP of, say, 3 percent and the 40 successive years of 4-percent efficiency improvements would reduce the economy's footprint only 37 percent by 2050. Finding enough new, clever, efficient technologies to reach the first-year 4 percent goal—to pick the "low-hanging fruit"—might be feasible. But by the twentieth or fortieth or sixtieth straight year, finding yet another 4 percent worth of technical fixes, not only to add to efficiency but also to address problems created by previous fixes, would present a formidable challenge; the addition of economic growth would put that fruit far out of reach.

In the past two decades, only two years—1991 and 2001—saw a decrease in the United States' greenhouse-gas emissions.[25] Both were years of economic recession. Situated comfortably between those recessions was the decade of the "weightless" economic revolution with its torrent of words and graphics, data and dollars, electrons and photons. For a few happy years, by slightly unfocusing one's eyes, it was possible to enjoy the illusion that capitalism could metamorphose into a clean and painless perpetual-motion machine. But all the while it was gobbling resources and spewing carbon faster than ever. Electrons and photons can only sell an SUV; they can't build one. The real economy—whether it's in the boom or bust part of the cycle—depends on the consumption of real stuff, and a lot of it. According to one source, 500 million computers, orphans of the dot-com boom, became obsolete in the United States between 1997 and 2007. They contained an estimated 6.3 billion pounds of plastic, 1.6 billion pounds of lead, and 632,000 pounds of mercury. Industry sources reported that 50 to 80 percent of US computer wastes collected for recycling are instead shipped off to Asia.[26]

A 2006 study looking back on the late 1990s found that the "weightless" boom of the late 1990s, like all booms, stimulated consumption of metals and energy—showing that efficiently deployed software can have the same impact as Jevons's efficiently burned coal. The author, Jonathan Perraton of the University of Sheffield, UK concluded:

> Thus, although the energy *intensity* of GDP has fallen in developed economies, total *use* of energy continues to rise. With rising absolute levels of energy use, economic expansion won't necessarily be freed from material constraints, even if new technologies offer the potential for continuous productivity growth[27] [his italics].

A remarkable study published in the journal *Energy* in 2004 considered what the overall effect would be if households in Sweden followed, en masse, the Swedish Environmental Agency's recommendations for "green consumption."[28] The Agency had recommended steps, well-supported by the scientific literature, toward a greener lifestyle: eating environmentally sound food from lower on the food chain, greater use of mass transportation, reduced car ownership, better fuel economy, less use of hot water and electricity in the home, and greater use of renewable energy sources. Adopting all of those good habits would clearly cut the household's expenses, but the researcher assumed that families would continue to spend their income at the same rate in the same overall patterns—only now they would make "greener choices" of products and services.

As expected, significant reductions in energy consumption and carbon dioxide emissions could be achieved by the year 2020 through individual actions like eating vegetarian rather than meat-based dinners or traveling more by train or bus and less by car. But doing those things also saved money, which was re-allocated to all the usual categories of spending in which a Swedish family typically engages—in the new green pattern wherever possible. Spending the money that was freed up by the green lifestyle brought the projected proportion of energy conserved by the Swedish population down to only 13 percent.

But that analysis assumed that the Swedes' real income would be stagnant over two decades. Even a modest 1-percent-per-year increase in real per capita income wiped out all gains in energy conservation, resulting in a small *increase* in energy consumption. A 2 percent annual income hike (closer to what economists would like to see) would lead to a 29 percent increase in energy use and a 13 percent increase in carbon dioxide emissions, despite 100 percent adherence to the recommended "green" lifestyle. Of course, people could choose not to spend the money they saved with green consumption. But unless they stuffed all the cash under their therapeutic Swedish mattresses and kept it there, that extra income would find its way into some part or other of the system and make trouble for the planet.

As we have seen, technological efficiency in food and medicine, including attempts to reduce environmental impact, end up stimulating growth, consumption, and waste. Another example of synergy between higher efficiency and harmful growth arises in business travel. Time spent in web-conferences and teleconferences has doubled since 2002, and spending is expected to continue rising, to $7 billion in 2010.[29] So that should be good news for the planet, right? As virtual meetings become more routine, it should be less necessary to ship live humans around the country and the globe at a huge economic and environmental cost. But it doesn't seem to work that way. US companies' penchant for travel has not been curbed by the communications revolution; rather, it seems to have been stimulated. In 2003, the most recent year for which there are complete records, companies sent their people on 210 million business trips.[30] Having gone into a slump after 9/11, business travel—about a third of all travel sales—has been growing at rates of more than 5 percent annually. And the number of companies using corporate or charter jets has shot up from 27 percent of all firms in 2002 to 56 percent in 2007.[31] According to Hubert Joly, president of industry giant Carson Wagonlit Travel, the increase in corporate travel is "a reflection of strong economic growth around the world and the globalization of the economy and corporations."[32]

Based on typical emissions estimates,[33] hauling a single economy-class passenger on a round trip across the United States releases almost two tons of carbon—about as much as would be burned to run the central air conditioning system back at that passenger's house for two and a half years. That's a lot, but high-end business travel generates even more: a first- or business-class passenger takes up more space on a plane (thereby accounting for a bigger share of its emissions), and corporate jets burn shocking amounts of fuel per person moved. Hotels, of course, are big energy hogs and waste factories. There is activity in the area of "green travel," but that will remain peripheral as long as the sheer volume of travel is going up instead of down.

An all-out effort in business to substitute fast-evolving communication technology for physical travel would seem to have huge potential to reduce ecological impact without serious sacrifice. But the internet, instead of substituting for wasteful travel, is boosting it. In efforts to maintain or increase their travel volume in the face of rising prices, companies are saving money by web-booking their tickets and hotel reservations. And some web-conferencing companies are awarding airline miles to their good customers![34]

One analyst told *USA Today* what he saw after 9/11: "From Sept. 15 to Dec. 15, 2001, we saw tremendous growth, people using teleconferencing as a substantial substitute for travel. When they got back to work on Jan. 1, they kissed their spouses goodbye, and headed for the nearest airport."[35] When business travel plummeted after 9/11, the world didn't come to an end (although the airlines used the occasion to rake in billions in government assistance). The host of ecological threats we face now—the biggest, of course, being rapid climate change—exceed anything a terrorist could concoct, and justify deep cutbacks in activities like business travel that the world can do without.

THE EUROPEAN MIRAGE

This book has drawn mostly on examples from the United States and India. Some might see that as distorting the message, if those countries represent extremes of capitalist enthusiasm. Many might point to Western Europe in particular as the home of greener, more fair capitalist economies. But the reality of Europe provides little encouragement.

The ecological footprint analysis that we examined in Chapter 1 for the United States showed that for the country to live within global biocapacity, the footprint per person would have to be reduced by 86 percent. Western Europe also would have to take a deep reduction, one of 74 percent, in its per-person footprint. Current efforts by the region's capitalist economies are falling far short of that because, as in the US, proposed environmental policies that would do significant harm to profits and economic growth are dead from the start.

In a review of Jeremy Rifkin's 2004 book *The European Dream: How Europe's Vision of the Future is Quietly Eclipsing the American Dream*,[36] Steve McGiffen, editor of the web publication *Spectrezine*, thoroughly eviscerated the idea that

Europe is blazing the trail toward a more economically fair and ecologically responsible future. Although the continent does indeed have some qualities that justify Rifkin's admiration—qualities like greater availability of locally produced food and shorter working hours for many—McGiffen argues that almost all of what's good about Europe is being attacked with great success by those now in power; that whatever vision its citizens might have for their continent, real economic power is held by multinational corporations and the European Union (EU) bureaucrats who back them up; that Europe is increasingly following the US economic model; that the EU exploits workers in its lower-wage member nations; that (using cheese as an example) the EU is "replacing varied local production with the unnecessary transport of mass-produced foodstuffs and other commodities, destroying small, artisanal producers and handing every advantage to the biggest corporations"; that shorter working hours and longer vacations are everywhere under attack—even in France, where "the 35-hour week no longer exists either as a meaningful, legally enforceable right or as an aspect of common practice"; and that the EU is constantly trying to force genetically modified crops down the throats of a populace that doesn't want them.[37] The European meat industry has earned praise for banning some cruel practices that are still legal in the US, but the fundamental cruelty of raising huge populations of animals in confinement is on the rise in both Western and Eastern Europe.

Europe's economic cast of characters resembles that in the US. Of the world's 25 largest corporations, 13 are European, with three oil companies and a car company holding the continent's top four slots.[38] The planet's three biggest pesticide makers and five of the ten biggest pharmaceutical companies are headquartered in Europe.[39] Wal-Mart may have become the biggest food retailer, but eight of the top ten chains are European, and a recent report says they have a poor overall record on diet and health.[40] The big push to privatize the world's water supplies is being led by the three dominant water corporations—two French and one British.[41]

Unlike the US, the EU has made attempts to adhere to the Kyoto protocol for curbing greenhouse-gas emissions. But to date, its market-friendly efforts have flopped spectacularly. In 2006, it was reported that the main European scheme for trading of carbon credits had started out by awarding excessively large allowances to big companies, who were able as a consequence to pump out greenhouse gases with ease while selling off unused credits at a huge financial gain; meanwhile, public institutions like schools and prisons were forced to spend their scarce funds to buy credits.[42] Then there was Kyoto's "Clean Development Mechanism" (CDM), which allowed wealthy nations to earn greenhouse credits by funding projects to reduce emissions in poorer nations. Credits are awarded according to the global-warming potential of the chemical compound being controlled; therefore, cutting a single ton of some refrigerants can earn as many credits as cutting 10,000 tons of carbon dioxide. That has helped European oil and motor-vehicle companies, for example, earn huge credits with cheap refrigerant-substitution programs while continuing to encourage fossil-fuel burning both at home and overseas.[43] Any carbon-credit trading scheme, when looked at from behind, is a way of canceling out necessary emissions cuts in some places by ensuring that damage can be continued or expanded in others.

Finally, events of the past few years have demonstrated that Europe has its share of out-and-out racists, plus many millions of more respectable folks who are eager to have access to the labor power of non-Europeans without having to deal with the non-Europeans themselves. Western Europe's economy would be nothing like what it is without the 5 million or so foreign guest workers who help build it but never experience much of the "European dream" themselves. Voters in country after country are showing a preference for business-friendly governments after having been failed by weak "left" governments. Even the more widely admired institutions and policies of today's Europe cannot be viewed as self-developed. They have been built partly on a foundation that includes the brutal colonialism of the past, continuing planet-wide exploitation of people and resources, Russian natural gas, and tacit reliance on the US military's dedication to keeping Persian Gulf and Central Asian petroleum flowing where the West wants it to flow.

Capitalism's defenders like to change the subject by conjuring up the ghosts of twentieth-century-style Communist societies, which were at least as environmentally destructive as their capitalist contemporaries. If future societies are to succeed in genuinely green planning where both capitalism and Communism failed, they must, in the words of ecosocialist Joel Kovel, allow people to "create and carry out those plans freely, not under orders from above. That is something the USSR never came close to achieving, and it was at the root of its ecological destructiveness. But to believe that the Soviet road is the only way to a post-capitalist society is to have no imagination."[44] After the fall of the Soviet Union, Russia's greenhouse emissions fell steeply and stayed low for years. That was the result of capitalist "reform," but not via the desired route: A deep, persistent economic depression, not enlightened green laws, held pollution in check.

Meanwhile, China has managed to bring together some of the worst features of capitalism and Communism in a single economy. Coal-burning spurred largely by China's full immersion into the world economy has had the following result, according to *Newsweek*:

> Virtually every day in December [2006], Beijing looked like a film negative of itself—spectral and acidic. And coal emissions do not respect borders. Sulfur dioxide discharges from China are being blown across the Pacific, causing acid rain in South Korea, Canada and Europe. Experts say that sulfur dioxide emissions from coal combustion ... kill about 400,000 Chinese prematurely annually.[45]

Even if China's efforts to employ "clean coal" technology succeed, economic growth is expected to increase the total amount of pollution.

In a review of Jared Diamond's 2005 bestseller *Collapse: How Societies Choose to Fail or Succeed*,[46] Richard Smith pointed out that in Diamond's prime examples of ecological collapse, the protagonists—Easter Islanders, Mayans, and Vikings of Greenland—all were brought down by "the ruling classes of those societies, who shut the rest of society out of decision-making and systematically made the 'wrong,' 'shortsighted' decisions that doomed their societies."[47] On the other hand, Diamond's ecological success stories, mostly from the Pacific Islands, were "small tribal village democracies where there were no distinctions of rank or class." Yet, argued Smith, when Diamond turns to our own situation, he forgets his own lesson, "ignores the systemic problems of capitalism that stand in the way of that needed radical change and instead falls back on the standard tried-and-failed strategy of lobbying, consumer boycotts, eco labeling, green marketing, asking corporations to adopt benign 'best practices,' and so on, the stock-in-trade strategy of the environmental lobbying industry...." There is no reason to expect such half-measures to succeed, because past examples, both good and bad, from whatever era or place we draw them, show that class exploitation and ecological devastation go hand-in-hand.

DIFFERENT KINDS OF IMPOSSIBILITY

It's not a pleasant prospect: A ubiquitous, seemingly unstoppable economic force that, if not stopped, will dash hopes of a decent life for everyone on a livable planet. The widely held belief that there's no longer any alternative to capitalism has imposed widespread paralysis at the very time that the likely catastrophic repercussions of further growth are rising into view. If the paralysis is to be overcome, the majority whom the system does not exist to serve, who don't live by and for accumulation of capital produced by the work of others, will have to proceed on two assumptions: that capitalism is not the natural and inevitable state of things, and that we are now living in a full world in which economic growth cannot be relied upon to provide for humanity. Addressing those who might despair of ever building a genuinely sustainable economy, Herman Daly has pointed out that

> one might well be tempted to declare that such a project would be impossible. But the alternative to a sustainable economy, an ever-growing economy, is biophysically impossible. In choosing between tackling a

political impossibility and a biophysical impossibility, I would judge the latter to be the more impossible and take my chances with the former.[48]

Instead of trying to repair or streamline a global economic system that refuses to be fixed, a small but growing ecosocialist movement is proposing ecologically viable alternatives. As outlined in Joel Kovel's book *The Enemy of Nature: The End of Capitalism or the End of the World*,[49] John Bellamy Foster's *Ecology Against Capitalism*[50] and *Vulnerable Planet: A Short Economic History of the Environment*,[51] and the pages of the journals *Monthly Review, Synthesis/Regeneration*, and *Capitalism Nature Socialism*, the ecosocialist position is rapidly evolving and far from monolithic. But it sits on a firm foundation. One ecosocialist manifesto states:

> Rationality limited by the capitalist market, with its short-sighted calculation of profit and loss, stands in intrinsic contradiction to ecological rationality, which takes into account the length of natural cycles. It is not a matter of contrasting "bad" ecocidal capitalists to "good" green capitalists; it is the system itself, based on ruthless competition, the demands of profitability, and the race for rapid profit, which is the destroyer of nature's balance. Would-be green capitalism is nothing but a publicity stunt, a label for the purpose of selling a commodity, or—in the best of cases—a local initiative equivalent to a drop of water on the arid soil of the capitalist desert.[52]

In the ecosocialist view, capitalism is working about as well as any pyramid scheme does before it goes bust. Statistics show clearly that the great wealth generated by huge increases in economic productivity in the US has gone mostly to employers and not workers; from 1973 to 2004, a 76 percent increase in output per hour worked resulted in only a 22 percent increase in real family income (and half of that entire income rise occurred during the 1995–2000 boom).[53] Productivity is greatly beloved by the investing class, because it's those additional increments of value produced by each worker that feed the accumulation of capital. Increased output per worker, accomplished either by improving the technology wielded by workers or by simply pushing them harder, doesn't imply greater resource efficiency; in fact, it almost always means increased use of resources and more wastes. If the fruits of that ramped-up productivity are going chiefly to those whose absolute wants were satisfied long ago and many times over, all that consumption and waste have accomplished nothing positive. When all consequences are accounted for, Marx's "general law of capitalist accumulation"—stating that there is always an "accumulation of misery, corresponding with an accumulation of capital"[54]—remains in effect. The green-capitalist promise of freedom from want and a healthy environment for all humanity through greater efficiency is as reliable as the always-lit neon sign at the local tavern promising "Free Beer Tomorrow."

Individual problems always suggest individual solutions: cracking down on Indian drug companies, for instance, or finding a substitute for a toxic chemical used in food packaging, going vegetarian or buying grass-fed beef. But if worldwide ecological disruption is a real probability, we have to solve these and a myriad other, more knotty problems simultaneously and without delay. That can't be done simply by implementing the right policies, one by one. Trying to curb consumption and waste simply by making it more expensive—through such direct charges or with a market in carbon credits or other licenses to pollute—will only amplify the advantage of the players with the deepest pockets. We can find a way to build a world in which people can live a decent life while keeping their ecological life-support system intact for future generations. But it won't be easy, and if the world's majority is compelled to spend scarce resources feeding a small minority's insatiable appetite for capital, it won't even be possible. Because any attempts to steer a capitalist economy toward eco-friendliness are guaranteed to reflect, for the most part, the interests of the economy's top decision makers, regulations imposed from above without the support of the majority of people are likely to be ignored or resisted. People's efforts to salvage the ecosphere will have to arise outside capitalist economics and outside the governments that serve capitalism's interests. Because of that, those efforts will be resisted as alien by economic elites.

An all-out push by the people of nations across the globe for worker ownership, green taxes (especially tough carbon taxes), stiff regulation of business, enforcement of antitrust laws, and redistribution of wealth would all be welcome developments, not only because of their direct effects but also because they would put pressure on the system. More good examples like those mentioned in the preceding chapters—the yet-to-be-built Green Health Center, Acción Fraterna and the community efforts in Anantapur, People's Grocery in Oakland—are needed, across every continent. But their goal must be to transcend the current economic order by refusing any longer to be ruled by a tiny class of owners, and that refusal will bring on terrible retaliation. Nevertheless, we have to work from the beginning to develop smaller organizations and structures that are specifically intended as parts of a system to succeed capitalism. And society's big institutions that Daly and others see as essential to sustainability—for limiting material throughput, population, and wealth disparities—must be intended from the beginning as parts of that successor system, not as appendages on a capitalist economy.

The first, highest obstacle to doing that is our continuing state of denial. Before *Homo sapiens* can start designing the kinds of local, regional, and world economies that are needed, we have to acknowledge and act on the fact that in the long run (which is getting shorter all the time) we cannot have both capitalism and a livable planet. To try to reverse the sequence, to say we can't give up on capitalism until we have a new system ready to drop into place will only give us more time to render the damage irreversible.

NOTES

All internet addresses cited were last accessed 24 June 2007.

PREFACE

1. H. Mooney, A. Cropper and W. Reid, "Confronting the human dilemma," *Nature* 434: 561–2 (2005).
2. Garret Keizer, "Climate, class, and claptrap," *Harper's*, June 2007.
3. Living on Earth, "Gore hits the global warming campaign trail," http://www.loe.org/shows/segments.htm?programID=07-P13-00003&segmentID=1
4. W. Norman and C. MacDonald, "Getting to the bottom of the 'triple bottom line'," *Business Ethics Quarterly* 14: 243–62 (2004).
5. J.B. Foster, "Capitalism and ecology: the nature of the contradiction," *Monthly Review* 54(4): 6–16 (2002).

CHAPTER 1

1. *Insurance Journal*, 27 March 2006.
2. *New York Times*, 27 March 2007.
3. *Minneapolis Star Tribune*, 20 December 2003.
4. *Chicago Tribune*, 11 February 2007.
5. US Preventive Service Task Force, Dept. of Health and Human Services: http://www.ahcpr.gov/clinic/cps3dix.htm. Ironically, as I was writing this chapter, I received a mail solicitation from Life Line Screening urging me to come to a local church for a one-day health fair: "You may think that your physician would order these screenings if they were necessary. However, insurance companies typically will not pay for screenings unless there are symptoms. Unfortunately, **50% of stroke victims have no symptoms**" (their emphasis).
6. Unmesh Kher, "The hospital wars," *Time*, 6 December 2006.
7. Penelope Patsuris, "Scan-dalous," *Forbes*, 27 July 2004.

8. A.W. Childs and E.D. Hunter, "Non-medical factors influencing use of diagnostic x-ray by physicians," *Medical Care* 10: 323–35 (1972).

9. US General Accounting Office, *Referrals to physician-owned imaging facilities warrant HCFA's scrutiny: report to the Chairman, Subcommittee on Health, Committee on Ways and Means, House of Representatives.* GAO/HEHS-95-2. Washington, DC: US General Accounting Office, 1994.

10. *New York Times*, 13 March 2004.

11. Patsuris, "Scan-dalous."

12. D.C. Levin and V.M. Rao, "Turf wars in radiology: the overutilization of imaging resulting from self-referral," *Journal of the American College of Radiology* 1: 169–72 (2004).

13. D. Studdert, M.M. Mello, W.M. Sage, C.M. DesRoches, J. Peugh, K. Zapert, T.A. Brennan, "Defensive medicine among high-risk specialist physicians in a volatile malpractice environment," *Journal of the American Medical Association (JAMA)* 293: 2609–17 (2005).

14. D. Merenstein, G. Daumit, N. Powe, "Use and costs of nonrecommended tests during routine preventive health exams," *American Journal of Preventive Medicine* 30: 521–7 (2006).

15. R.J.H. Hammett and R.D. Harris, "Halting the growth in diagnostic testing," *Medical Journal of Australia* 177: 124–5 (2002).

16. L.E. Boulware, B.G. Jaar, M.E. Tarver-Carr, F.L. Brancati, and N.R. Powe, "Screening for proteinuria in U.S. adults: a cost-effectiveness analysis," *JAMA* 290: 3101–14 (2003) and Merenstein et al., "Use and costs."

17. Merenstein et al., "Use and costs."

18. CBS News, 19 May 2006, http://www.cbsnews.com/stories/2006/05/19/health/webmd/main1637144.shtml

19. Kher, "The hospital wars."

20. *Washington Post*, 14 February 2007.

21. National Academy of Sciences, *To Err Is Human: Building a Safer Health System*, 2000, p. 1 (text online at http://www.nap.edu/books/0309068371/html/).

22. R.L. Howard, A.J. Avery, S. Slavenburg, S. Royal, G. Pipe, P. Lucassen, and M. Pirmohamed, "Which drugs cause preventable admissions to hospital? A systematic review," *British Journal of Clinical Pharmacology* 63: 136–47 (2006).

23. S.H. Woolf, "Potential health and economic consequences of misplaced priorities," *JAMA* 297: 523–6 (2007).

24. D. Carpenter, "The boom goes on," *Trustee* 59(3): 6–10 (March, 2006).

25. Office of Technology Assessment, US Congress, "Issues in medical waste management—background paper," OTA-BP-O-49, Washington, DC: US Government Printing Office, 1988.

26. Jessica Pierce and Andrew Jameton, *The Ethics of Environmentally Responsible Health Care*, Oxford: Oxford University Press, 2004, p. 10.

27. Ibid., p. 11.

28. McKibben, T. Horan, J.I. Tokars, G. Fowler, D.M. Cardo, M.L. Pearson, and P.J. Brennan, "Guidance on public reporting of healthcare-associated infections: recommendations of the healthcare infection control practices advisory committee," *American Journal of Infection Control* 33: 217–26 (2005).

29. Andrew Jameton, "Is a modest health care system possible?" *Synthesis/Regeneration*, Fall 2007.

30. Interview with Andrew Jameton, February 2006.

31. Kaiser Family Foundation, "Trends and indicators in the changing health care marketplace," Publication number 7031 (2005); http://www.kff.org/insurance/7031/index.cfm

32. Assistant Secretary for Planning and Evaluation, US Dept. of Health and Human Services, "Long-term growth of medical expenditures—public and private," May 2005, http://aspe.hhs.gov/health/MedicalExpenditures/ib.pdf

33. P.B. Ginsburg, B.C. Strunk, M.I. Banker, and J.P. Cookson, "Tracking health care costs: continued stability but at high rates in 2005," *Health Affairs* 25: w486–w495. (2006) web only: http://content.healthaffairs.org/cgi/content/abstract/hlthaff.25.w486v1

34. At the 2007 G8 summit of wealthy nations, Nicolas Sarkozy, the newly elected President of France, used this common phrase in an especially revealing way, stating that "we want respect for the planet at the same time because if we destroy the planet there will be growth for nobody," *Religious Intelligence*, 8 June 2007, http://www.religiousintelligence.co.uk/news/?NewsID=807

35. Redefining Progress, "2005 Ecological Footprint of Nations," http://www.rprogress.org/publications/2006/Footprint%20of%20Nations%202005.pdf

36. The Board of the Millennium Ecosystem Assessment, "Living beyond our means: natural assets and human well-being," March 2005; http://ma.caudillweb.com/proxy/document.429.aspx

37. Millennium Ecosystem Assessment, *Ecosystems and human well-being: policy responses*, Washington, DC: Island Press, 2005, p. 381.

38. For a thorough demolition of the Sierra Club's prediction that renewable energy will be sufficient, see Don Fitz, "Consume,

consume, consume," *Synthesis/Regeneration*, Summer 2007. True renewability means forgetting grain ethanol and nuclear power, for example, while accounting for the vast expenditure of nonrenewable resources that will be required to build a truly renewable infrastructure. And efforts to derive anything close to 100 percent of energy from truly renewable sources will run up against inevitable, irresistible economic pressure to exploit available fossil fuels.

39. Michael Mandel and Joseph Weber, "What's really propping up the economy," *Business Week*, 25 September 2006.
40. Russ Brit, "Health insurers getting bigger cut of medical dollars," *Investors Business Daily*, 15 October 2004; http://investors.com/breakingnews.asp?journalid=23544168&brk=1
41. Kaiser Family Foundation, "Trends and indicators."
42. Donald Bartlett and James Steele, *Critical Condition: How Health Care in America Became Big Business & Bad Medicine*, New York: Doubleday, 2004.
43. Ibid., pp. 83–4.
44. Ibid., p. 86.
45. *Washington Post*, 3 July 2005.
46. Bartlett and Steele, *Critical Condition*, p. 102.
47. Hospitals for a Healthy Environment, www.h2e-online.org/pubs/mercuryreport.pdf
48. Health Care Without Harm; http://www.noharm.org/us/pvcDehp/reducingPVC#ordinances
49. Green Guide for Health Care; http://www.gghc.org
50. Telephone interview with Ted Schettler, February 2006.
51. Pierce and Jameton, *The Ethics of Environmentally Responsible Health Care*.
52. D.V. Canyon, "Saving lives or saving the planet: the ethics of environmentally responsible health care," *The Lancet* 363: 9425 (2004).
53. Telephone interview with Jessica Pierce, February 2006.
54. Jameton, "Is a modest health care system possible?"
55. G. Salkeld, D. Henry, S. Hill, D. Lang, and N. Freemantle, "What drives healthcare spending priorities? An international survey of health-care professionals," *Public Library of Science, Medicine (PLoS Med)* 4: e94 (2007).
56. R.W. Bush, "Reducing waste in US health care systems," *JAMA* 297: 871–4 (2007).
57. Bartlett and Steele, *Critical Condition*, p. 114.
58. Mandel and Weber, "What's really propping up the economy."
59. Friends of the Earth UK, "More isn't always better: a special briefing on growth and quality of life in the UK," http://www.foe.co.uk/resource/briefings/more_isnt_better.pdf

60. P.A. Lawn, "A theoretical foundation to support the Index of Sustainable Economic Welfare (ISEW), Genuine Progress Indicator (GPI), and other related indexes," *Ecological Economics* 44:105–18 (2003).
61. John Talberth, Clifford Cobb, and Noah Slattery, "The Genuine Progress Indicator 2006: a tool for sustainable development," http://www.rprogress.org/newpubs/2007/GPI%202006.pdf

CHAPTER 2

1. Rich Thomaselli, "Ten years later: direct to consumer drug advertising," *Advertising Age*, 1 October 2006.
2. K.S. Fink and P.J. Byrns, "Changing prescribing patterns and increasing prescription expenditures in Medicaid," *Annals of Family Medicine* 2: 488–93 (2004).
3. Marcia Angell, "The truth about the drug companies," *New York Review of Books*, 15 July 2004.
4. Following each condition is given the prevalence in adults claimed by industry and the source of the figure: restless legs syndrome, 10 percent, *PLoS Medicine* 3(4): e170; irritable bowel syndrome, 20 percent, Moynihan and Cassels, *Selling Sickness* (see below); bipolar disorder, 5 percent, *PLoS Medicine* 3(4): e185; attention deficit disorder, 4 percent, *Selling Sickness*; social anxiety disorder, 13 percent, *Selling Sickness*; erectile dysfunction, 52 percent, *PLoS Medicine* 3(4): e132; female sexual dysfunction, 43 percent, *BMJ* [*British Medical Journal*] 326: 45–7 (2003); sleep disorders, 75, *New Scientist*, 18 February 2006. I took the proportions of people who would *not* have each individual condition and, assuming that your chances of having one of these conditions does not affect your chances of having another, multiplied all of the appropriate percentages together for the conditions appropriate to each sex. The result was that only 7 percent of men and 8 percent of women would have none of the conditions. None of the figures given for individual diseases is at all precise, but based on those figures, more than 90 percent of adults would probably suffer from at least one.
5. Lynn Payer, *Disease Mongers: How Doctors, Drug Companies, and Insurers Are Making You Feel Sick*, New York: John Wiley & Sons, 1992.
6. Ray Moynihan and Alan Cassels, *Selling Sickness: How The World's Biggest Pharmaceutical Companies Are Turning Us All Into Patients*, Vancouver: Greystone Books, 2005.

7. S. Woloshin and L.M. Schwartz, "Giving legs to restless legs: a case study of how the media helps make people sick," *PLoS Medicine* 3(4): e170 (2006).

8. *Guardian*, 28 April 2006.

9. Telephone interview with David Henry, May 2006.

10. Moynihan and Cassels, *Selling Sickness*, pp. 156–74.

11. I. Schwetz, S. Bradesi, and E.A. Mayer, "Current insights into the pathophysiology of irritable bowel syndrome," *Current Gastroenterology Reports* 5: 331–6 (2003).

12. Moynihan and Cassels, *Selling Sickness*, pp. 168–9.

13. Ibid., p. 160.

14. Lists of all Bitter Pill Award recipients, and information on the FDA letter, may be viewed at http://www.bitterpillawards.org/

15. http://www.rxawards.com/pastawardsub06.html

16. http://www.help4adhd.org/en/treatment/medical/WWK3

17. J. Lexchin, "Bigger and better: how Pfizer redefined erectile dysfunction," *PLoS Medicine* 3(4): e132 (2006).

18. L. Tiefer, "Female sexual dysfunction: a case study of disease mongering and activist resistance," *PLoS Medicine* 3(4): e178 (2006).

19. D. Healy, "The latest mania: selling bipolar disorder," *PLoS Med* 3(4): e185 (2006).

20. E.M. Hunkeler, B. Fireman, J. Lee, R. Diamond, J. Hamilton, C.X. He, R. Dea, W.B. Nowell, and W.A. Hargreaves, "Trends in use of antidepressants, lithium, and anticonvulsants in Kaiser Permanente-insured youths, 1994–2003," *Journal of Child and Adolescent Psychopharmacology* 15: 26–37 (2005).

21. Moynihan and Cassels, *Selling Sickness*, pp. 119–38.

22. D. Josefson, "Cholesterol guidelines will triple numbers taking drugs," *BMJ* 322: 1270 (2001).

23. NIH press release, 12 July 2004; http://www.nih.gov/news/pr/jul2004/nhlbi-12.htm

24. Jeanne Lenzer, "US consumer body calls for review of cholesterol guidelines," *BMJ* 329: 759 (2004).

25. T.J. Wilkin and D. Devendra, "Bone densitometry is not a good predictor of hip fracture," *BMJ* 323: 795–7 (2001).

26. Harvey R. Colten and Bruce M. Altevogt (eds), *Sleep Disorders and Sleep Deprivation: An Unmet Public Health Problem*, Washington, DC: National Academies Press, 2006.

27. Telephone interview with David Henry, May 2006.

28. Commercial Alert press release, 24 May 2006; http://www.commondreams.org/news2006/0524-08.htm

29. *Advertising Age*, 16 August 2006; http://www.emarketer.com/Article. aspx?1004120
30. T.A. Brennan, D.J. Rothman, L. Blank, D. Blumenthal, S.C. Chimonas, J.J. Cohen, J. Goldman, J.P. Kassirer, H. Kimball, J. Naughton, and N. Smelser, "Health industry practices that create conflicts of interest: a policy proposal for academic medical centers," *JAMA* 295: 429–33 (2006).
31. "The labyrinth of care" is the heading of Chapter 4 in Bartlett and Steele, *Critical Condition*.
32. This and succeeding quotes and information from Kathleen Slattery-Moschkau are from telephone interviews conducted in May 2007.
33. See A. Wazana, "Physicians and the pharmaceutical industry: is a gift ever just a gift?" *JAMA* 283: 373–80 (2000). Slattery Moschkau told me: "Most weeks, I'd bring four to five lunches into clinics. And a couple of breakfasts: bagels, good coffee. If it was really hot weather, I'd bring ice cream. The idea was just to get them to stop and talk. Most would be too gracious just to take some food and walk away. Some did do that, but most would think, 'I've gotta give her a minute.' More clinics are shutting out food these days. But the arms race still goes on. Every year, there are more reps all trying to get time with the same doctors." For a while, to get in a word with their targets, Slattery-Moschkau's and other reps resorted to a tactic called "dine and dash": "I'd tell a doctor, look, if you stop at such-and-such a restaurant on your way home, you can get a take-home meal for your family, on us. Then I could meet the doctor at the restaurant and we could talk while he was waiting for his order."
34. *Rutland* (Vermont) *Herald*, 12 June 2007.
35. Wazana, "Physicians and the pharmaceutical industry."
36. http://www.pharmrep.com/pharmrep/article/articleDetail. jsp?id=114445
37. T.A. Hemphill, "Physicians and the pharmaceutical industry: a reappraisal of marketing codes of conduct," *Business and Society Review* 111: 323–36 (2006).
38. Brennan et al., "Health industry practices."
39. R.F. Wright and W.J. Lundstrom, "Physicians' perceptions of pharmaceutical sales representatives: a model for analysing the customer relationship," *Journal of Medical Marketing* 4: 29–38 (2004).

CHAPTER 3

1. Nicholas Georgescu-Roegen, *The Entropy Law and the Economic Process*, New York: Harvard University Press, 1971, pp. 305–6.
2. *Washington Post*, 17 June 2007.

3. KPMG International, "The Indian pharmaceutical industry: collaboration for growth" (2006); http://www.kpmg.fi/Binary.aspx? Section=174&Item=2888
4. Ibid.
5. *The Hindu*, 3 July 2006.
6. Government of India, "National draft pharmaceuticals policy, 2006, Part A" (2005); http://www.usaindiachamber.org/images/Draft%20 Pharma%20Poicy_2005.pdf
7. This chemical should not be confused with methyl isocyanate, which poisoned and killed tens of thousands in Bhopal, India in 1984. Methyl isothiocyanate is a different compound and is not as acutely toxic.
8. Interview with Dr. Allani Kishan Rao, January 2005.
9. *Washington Post*, 17 June 2007.
10. Greenpeace India, "State of community health at Medak District, 2004," http://www.greenpeace.org/raw/content/india/press/reports/ state-of-community-health-at-m.pdf
11. G.A. Rao, M. Parabrahman, M. Haribabu, S. Bapu Rao, and A.S. Raj, "Report of fact finding committee, constituted by the Hon'ble High Court in its order dated 25th September 2003 in W.P. No. 19661/02," March, 2004. Photocopy of the original report, which I borrowed from the Andhra Pradesh Pollution Control Board in February, 2005. An official allowed me to take the report across the street to a copy shop.
12. To obtain this figure, I compared the Fact Finding Committee's results with an international standard of no more than 1,000 milligrams per liter total dissolved solids, cited in a 2000 report by scientists from the National Geophysical Research Institute: V.V.S. Gurunadha Rao, R.L. Dhar, and K. Subramanyam, "Assessment of contaminant migration in groundwater from an industrial development area, Medak District, Andhra Pradesh, India," *Water, Air, and Soil Pollution* 128: 369–89 (2001).
13. KPMG International, "The Indian pharmaceutical industry."
14. Ibid.
15. Almost everyone I spoke with, on both sides of the dispute and in the middle, mentioned a report by a committee of experts appointed by the High Court to study the situation in 2002–3, and leaned on it for support. But no one I spoke with had a copy of the report or had ever even seen a copy. Nicholas Piramal executives told me the report proves that the village's water is no longer polluted, and that it recommends that the Court award no compensation to farmers. But they had not actually seen the report. Officials at the local Pollution Control Board office said they don't have a copy, and more senior

officials at the Board's headquarters couldn't provide one either. An attorney for the villagers had requested a copy, but never got one. And, of course, no one in the village had seen the report. One farmer told me, "People come all the time and take water samples, but they never come back to show us any results." See Stan Cox, "Fighting Big Pharma in little Digwal," *CounterPunch.org*, 15 February 2005; http://www.counterpunch.org/cox02152005.html

16. G.A. Rao et al., "Report of fact finding committee."
17. Interview with Venkat Ram, December 2004.
18. SCMC letter—Letter from G. Thyagarajan, Chairman, Supreme Court Monitoring Committee to Dr. Mohan Kanda, Chairman, Andhra Pradesh Pollution Control Board, 11 February 2005; http://www.scmc.info/communications/scmc_to_ap_government.htm
19. Supreme Court Monitoring Committee, "Report on AP compliance with apex court order dated 14.10.2003 and SCMC directions," 13 March 2006; http://www.scmc.info/reports/andhrapradesh/report_march_8_2006.html
20. Intergovernmental Panel on Climate Change, Working Group II, "Climate change 2007: impacts, adaptation and vulnerability: summary for policymakers," 8 March 2006; http://www.ipcc-wg2.org/index.html
21. Interview with Dr. S. Jeevananda Reddy, February 2007.
22. Interview with Dr. Allani Kishan Rao, February 2007.
23. Michael Rosen, "Indian trade exports boggle the mind," *Wisconsin Technology Network*, 26 March 2007; http://wistechnology.com/article.php?id=3803
24. Confidential interview with a Hyderabad resident connected to one of the study's authors, January 2007.

CHAPTER 4

1. Stan Cox and Marty Bender, "Warning—this diet is not for everyone: The Atkins diet's ecological side effects," in Lisa Heldke, Kerri Mommer, and Cynthia Pineo (eds), *The Atkins Diet and Philosophy*, Chicago: Open Court, 2005, pp. 170–81. Marty died of cancer in the spring of 2006, and he is badly missed. And were he still with us, this chapter would be much better.
2. In our report, we used the term "overweight" in the interest of brevity but reluctantly, to indicate a state of being heavier than the standard range for one's height. This is a highly controversial area, of course, and we were not endorsing the idea that society should dictate desirable body size; that is each person's choice.

Most of the time here, I refer to "people who want to lose weight," "dieters," etc.

3. Immanuel Kant, *Groundwork for the Metaphysics of Morals*. New Haven: Yale University Press, 2002, p. 14.

4. *World Watch*, July/August 2004.

5. Gary Gardner and Brian Halweil, "Overfed and underfed: The global epidemic of malnutrition," Worldwatch paper #150, Washington, DC, Worldwatch Institute.

6. The calculations underlying these figures are summarized in Cox and Bender, "Warning" and in Cox and Bender, "How low-carb should you go?," Grist Magazine, 9 February 2004, http://www.grist.org/comments/soapbox/2004/02/09/you. After publication of that article, I received a few emails from people who had followed Atkins's nutritional guidelines but without eating meat. But it's important to realize that a low-carb vegetarian diet would never have taken the nation and world by storm the way Atkins did. The prospect of eating tasty meat at every meal was an important part of the diet's appeal.

7. P. Walker, P. Rhubart-Berg, S. McKenzie, K. Kelling, and R.S. Lawrence, "Public health implications of meat production and consumption," *Public Health Nutrition* 8: 348–56 (2005).

8. Jerry D. Glover, Cindy M. Cox, and John P. Reganold, "Future farming: a return to roots?," *Scientific American*, August 2007.

9. Bar On, Bat-Ami, "Commodious diets, or could a Marxist do Atkins?," in *The Atkins Diet and Philosophy*, pp. 125–35.

10. Karen Hudgins, "Low-carb losses," *Fiscal Notes*, Office of the Texas Comptroller of Public Accounts, March, 2005; http://www.cpa.state.tx.us/comptrol/fnotes/fn0503/low-carb.html

11. KCCI-TV, Des Moines, Iowa, "Are low-carb diets boosting hog production?," 28 March 2004; http://www.kcci.com/money/2955442/detail.html

12. CarbWire, "Smithfield Foods' profits up, carbs cited," 9 June 2004; http://www.carbwire.com/2004/06/09/pork-sales-up

13. C.D. Gardner, A. Kiazand, S. Alhassan, S. Kim, R.S. Stafford, R.R. Balise, H.C. Kraemer, and A.C. King, "Comparison of the Atkins, Zone, Ornish, and LEARN diets for change in weight and related risk factors among overweight premenopausal women," *JAMA* 297: 969–77 (2007).

14. Reuters, "Tyson on Atkins: 'Great News'", 7 March 2007; http://www.reuters.com/news/video/videoStory?videoId=20288

15. Marketdata Enterprises press release, 19 April 2007; http://www.prwebdirect.com/releases/2007/4/prweb520127.htm

16. Graham Greene, *The Comedians*, London: Penguin Books, 1965, p. 168.
17. NIH Technology Assessment Conference Panel, "Methods for voluntary weight loss and control," *Annals of Internal Medicine* 119: 764–70 (1993).
18. Aravinda Meera Guntupalli, "Inquiry into the simultaneous existence of malnutrition and overweight in India," presentation to the Population Association of America annual meeting, 31 March 2006; http://paa2006.princeton.edu/download.aspx?submissionId= 61837
19. G. Mudur, "Asia grapples with obesity epidemics," *BMJ* 326: 515 (2003).
20. M.A. Mendez, C.A. Monteiro, and B.M. Popkin, "Overweight exceeds underweight among women in most developing countries," *American Journal of Clinical Nutrition* 81: 714–21 (2005).
21. Mudur, "Asia grapples with obesity epidemics."
22. Guntupalli, "Inquiry."
23. MSNBC, 10 October 2006; http://www.msnbc.msn.com/id/14604784/
24. George Monbiot, "Mass medication with Omega 3 would wipe out global fish stocks," *Guardian*, 20 June 2006.
25. Charles Clover, *The End of the Line: How Over-fishing is Changing the World and What We Eat*, London: Ebury Press, 2004.
26. National Oceanic and Atmospheric Administration, Fisheries Statistics Division, "Processed Fishery Products," http://www.st.nmfs.gov/st1/fus/fus03/05_process2003.pdf and http://www.st.nmfs.gov/st1/fus/fus04/05_process2004.pdf
27. B. Worm, E.B. Barbier, N. Beaumont, J.E. Duffy, C. Folke, B.S. Halpern, J.B.C. Jackson, H.K. Lotze, F. Micheli, S.R. Palumbi, E. Sala, K.A. Selkoe, J.J. Stachowicz, and R. Watson, "Impacts of biodiversity loss on ocean ecosystem services," *Science* 314: 787–90 (2006).
28. U.N. Das, "Essential fatty acids: biochemistry, physiology and pathology," *Biotechnology Journal* 1: 420–39 (2006).
29. K. Nuernberg, G. Nuernberg, K. Endera, S. Lorenza, K. Winkler, R. Rickert, and H. Steinhart, "N-3 fatty acids and conjugated linoleic acids of longissimus muscle in beef cattle," *European Journal of Lipid Science and Technology* 104: 463–71 (2002).
30. Associated Press, 9 May 2006.
31. Tea Council of the USA; http://www.teausa.com/general/fda_decline.cfm
32. Andy Isaacson, "Steeped in tea," *Utne Reader*, January/February 2007.

33. http://www.teatreasury.com/health.htm
34. Isaacson, "Steeped in tea." Of course, a wide variety of herbal teas are sold, but the foundation of the market remains the "real" thing from the tea plant *Camellia sinensis*.
35. Damien McElroy, "Unsafe pesticide and lead levels: that's China to a tea," *The Age* (Australia), 15 June 2002, and National Geographic Society, "The green guide product report: tea," http://www.thegreenguide.com/reports/product.mhtml?id=56
36. Email interview with Suprabha Seshan, June 2007. Seshan has also seen attempts to help farmers end up doing even more harm: "Then there is the economics of it. Can the global market deliver its promises and keep them? We have seen a series of failures one after the other because of the tie-up with global forces, where before there was a nationalized board that standardized prices. The price of fresh leaf tea fell from 22 rupees to 2 rupees per kilo in a period of three years after a massive government-subsidized programme that got thousands of small farmers to convert to tea and then they were abandoned! That led to suicides in a relatively rich part of the country."
37. S. Biswas, D. Chokraborty, S. Berman, and J. Berman, "Nutritional survey of tea workers on closed, re-opened, and open tea plantations of the Dooars region, West Bengal, India, October, 2005," West Bengal Agricultural Workers' Association, 2006, http://www.iufdocuments.org/www/documents/AJWSnutritionreport.pdf
38. R.A. Myers and B. Worm, "Extinction, survival or recovery of large predatory fishes," *Philosophical Transactions of the Royal Society* B 360: 13–20 (2005).
39. R.A. Myers, J.K. Baum, T.D. Shepherd, S.P. Powers, and C.H. Peterson, "Cascading effects of the loss of apex predatory sharks from a coastal ocean," *Science* 315: 1846–50 (2007).
40. H. Pellissier, "Shark fin soup: an eco-catastrophe?," *San Francisco Chronicle*, 20 January 2003.
41. Peggy Nauts, "Koi Palace," *San Francisco* magazine, 2007; http://www.sanfranmag.com/archives/view_story/451/
42. British Columbia Cancer Agency, 2000; http://www.bccancer.bc.ca/PPI/UnconventionalTherapies/SharkCartilageCartilateCartiladeBenefin AE941Neovastat.htm
43. William Lane and Linda Comac, *Sharks Don't Get Cancer: How Shark Cartilage Could Save Your Life*, New York: Avery, 1992.
44. G.K. Ostrander, K.C. Cheng, J.C. Wolf, and M.J. Wolfe, "Shark cartilage, cancer and the growing threat of pseudoscience," *Cancer Research* 64: 8485–91 (2004).
45. FDA press release, 13 July 2004; http://www.fda.gov/bbs/topics/news/2004/NEW01086.html

46. *Los Angeles Times*, 26 December 2006.
47. *The Star* (South Africa), 27 November 2006.
48. Ibid.
49. *Los Angeles Times*, 26 December 2006.
50. See, for example, Emily Arnold and Janet Larsen, "Bottled water: pouring resources down the drain," Earth Policy Institute, 2 February 2006; http://www.earth-policy.org/Updates/2006/Update51.htm
51. A. Fenwick, "Waterborne infectious diseases—could they be consigned to history?," *Science* 313: 1077–81 (2006).
52. J. Tibbetts, "Water World 2000," *Environmental Health Perspectives* 108: A69–A73 (2000).
53. *San Francisco Chronicle*, 17 January 2007.

CHAPTER 5

1. Nicholas Georgescu-Roegen, *The Entropy Law and the Economic Process*, New York: Harvard University Press, 1971, pp. 250–1.
2. http://today.msnbc.msn.com/id/6644980
3. Howard Elitzak, "Food Marketing Costs: A 1990's Retrospective," *FoodReview* 23: 27–30 (2000).
4. Data were obtained from the US Bureau of Economic Analysis; http://bea.gov/bea/dn2/gdpbyind_data.htm
5. Data were obtained from the Economic Research Service of the US Dept. of Agriculture; http://www.ers.usda.gov/Data/FarmIncome/finfidmu.htm
6. R. Allen, G. Harris, and K. Young, "National Agricultural Statistics Service, Economic Statistics And Information Resources Committee Report, July 2003 through June 2004," http://www4.ncsu.edu/~bkgoodwi/esirc/NASS.pdf
7. W.C. Lowdermilk, "Conquest of the land through 7,000 years," Agricultural Information Bulletin 99. USDA-Soil Conservation Service, Washington DC (1975) and Daniel Hillel, *Out of the Earth: Civilization and the Life of the Soil*, Berkeley, California: University of California Press, 1992.
8. John Bellamy Foster and Fred Magdoff, "Leibig, Marx, and the depletion of soil fertility: relevance for today's agriculture," in *Hungry for Profit*, Fred Magdoff, John Bellamy Foster, and Frederick Buttel (eds), New York: Monthly Review Press, 2000.
9. Wes Jackson, *New Roots for Agriculture*, Lincoln, Nebraska: University of Nebraska Press, 1980.
10. US Government Accountability Office, "Homeland security: much is being done to protect agriculture from a terrorist attack, but

important challenges remain," GAO report number GAO-05-214 (2005).

11. US Centers for Disease Control and Prevention, "Fact sheet: foodborne illness," http://www.cdc.gov/ncidod/dbmd/diseaseinfo/foodborneinfections_g.htm

12. Partnership for Food Safety Education, "Least wanted foodborne pathogens," http://www.fightbac.org/content/view/14/21

13. P. Walker, P. Rhubart-Berg, S. McKenzie, K. Kelling, and R.S. Lawrence, "Public health implications of meat production and consumption," *Public Health Nutrition* 8: 348–56 (2005).

14. J.B. Russell, F. Diez-Gonzalez, and G.N. Jarvis, "Effects of diet shifts on *Escherichia coli* in cattle," *Journal of Dairy Science* 83: 863–73 (2000).

15. Kansas Department of Health and Environment, "2004 Kansas water quality assessment (305(b) report)," http://www.kdheks.gov/befs/download/305b04text12f.pdf

16. US Census Bureau, "The 2007 statistical abstract," http://www.census.gov/compendia/statab/agriculture/meat_and_livestock

17. Environmental Defense, Scorecard Pollution Information Site; http://www.scorecard.org/env-releases/aw/us.tcl

18. "Census of Agriculture—Volume 1."

19. R.L. Kellogg, C.H. Lander, D.C. Moffitt, and N. Gollehon, "Manure nutrients relative to the capacity of cropland and pastureland to assimilate nutrients: spatial and temporal trends for the United States," Natural Resources Conservation Service, US Dept. of Agriculture, publication number nps00-0579 (2000).

20. N.S. Simon, O.P. Bricker, W. Newell, J. McCoy, and R. Morawe, "The distribution of phosphorus in Popes Creek, VA, and in the Pocomoke River, MD: two watersheds with different land management practices in the Chesapeake Bay basin," *Journal of Water, Air, & Soil Pollution* 164: 189–204 (2005).

21. F.M. Mitloehner and M.B. Schenker, "Environmental exposure and health effects from concentrated animal feeding operations," *Epidemiology* 18: 309–11 (2007).

22. S.T. Sigurdarson and J.N. Kline, "School proximity to concentrated animal feeding operations and prevalence of asthma in students," *Chest* 129: 1486–91 (2006).

23. J. Zhu and X. Li, "A field study on downwind odor transport from swine facilities," *Journal of Environmental Science and Health* B35: 245–58 (2000).

24. S.S. Schiffman, C.E. Studwell, L.R. Landerman, K. Berman, and J.S. Sundy, "Symptomatic effects of exposure to diluted air sampled from a swine confinement atmosphere on healthy human subjects,"

Environmental Health Perspectives 113: 567–76 (2005) and R.C. Avery, S. Wing, S.W. Marshall, and S.S. Schiffman, "Odor from industrial hog farming operations and mucosal immune function in neighbors," *Archives of Environmental Health* 59: 101–8 (2004).

25. P. Walker et al., "Public health implications of meat."
26. A. Chapin, A. Rule, K. Gibson, T. Buckley, and K. Schwab, "Airborne multi-drug resistant bacteria isolated from a concentrated swine feeding operation," *Environmental Health Perspectives* 113: 137–42 (2005) and A.R. Sapkota, K.K. Ojo, M.C. Roberts, K.J. Schwab, "Antibiotic resistance genes in multidrug-resistant *Enterococcus* spp. and *Streptococcus* spp. recovered from the indoor air of a large-scale swine-feeding operation," *Letters in Applied Microbiology* 43: 534–40 (2006).
27. A.M. France, C.F. Marrs, L. Zhang, and B. Foxman, "Reply to Riley and Manges and to Johnson," *Clinical Infectious Diseases* 41: 568–70 (2005).
28. M. Ramchandani, A.R. Manges, C. DebRoy, S.P. Smith, J.R. Johnson, and L.W. Riley, "Possible animal origin of human-associated, multidrug-resistant, uropathogenic *Escherichia coli*," *Clinical Infectious Diseases* 40: 251–7 (2005).
29. *Washington Post*, 24 September 2001.
30. KRSNetwork, "U.S. pesticide industry report: executive summary" (2005); http://www.knowtify.net/2005USPestIndReptExecSum.pdf
31. A. Blair, D. Sandler, K. Thomas, J. Hoppin, F. Kamel, J. Coble, W. Lee, J. Rusiecki, C. Knott, M. Dosemeci, C.F. Lynch, J. Lubin, and M. Alavanja, "Disease and injury among participants in the Agricultural Health Study," *Journal of Agricultural Safety and Health* 11: 141–50 (2005).
32. Project leader Dr. Aaron Blair quoted in Stan Cox, "Turf Wars," AlterNet, 17 November 2005; http://www.alternet.org/environment/28361
33. A. Ascherio, H. Chen, M.G. Weisskopf, E. O'Reilly, M.L. McCullough, E.E. Calle, M.A. Schwarzschild, and M.J. Thun, "Pesticide exposure and risk for Parkinson's disease," *Annals of Neurology* 60: 197–203 (2006).
34. R. Das, A. Steege, S. Baron, J. Beckman, and R. Harrison, "Pesticide-related illness among migrant farm workers in the United States," *International Journal of Occupational and Environmental Health* 7: 303–12 (2001).
35. D. Villarejo, "The health of U.S. hired farmworkers," *Annual Review of Public Health* 24: 175–93 (2003).

36. Centers for Disease Control and Prevention, "Third national report on human exposure to environmental chemicals," (2005); http://www.cdc.gov/exposurereport

37. *Time*, 6 December 1968.

38. The Monsanto Company, 2006 Monsanto technology/stewardship agreement. Document image can be seen at http://www.farmsource.com/images/pdf/2006%20EMTA%20Rev3.pdf

39. Andrew Kimbrell and Joseph Mendelson, *Monsanto vs. U.S. Farmers*, Washington, DC: Center for Food Safety, 2005.

40. Phone number dialed 30 May 2007.

41. Kimbrell and Mendelson, *Monsanto vs. U.S. Farmers*.

42. Lisa M. Hamilton, "Draining the Gene Pool," The Nation (online only), 4 December 2006; http://www.thenation.com/doc/20061218/hamilton; and Stan Cox, 'The Gene Rush," AlterNet, 14 December 2005; http://www.alternet.org/story/29532/

43. W.I. Thomas, "Transferring the Ga^s factor for dent-incompatibility to dent-compatible lines of popcorn," *Agronomy Journal* 47: 440–1 (1955).

44. Syngenta press release, 20 February 2004; http://www.syngenta.com/en/media/article.aspx?pr=022004_2&Lang=en

45. Karl Marx, *Capital: an Abridged Edition*, David McClellan (ed.), Oxford, UK: Oxford University Press, 1995, p. 298.

46. US Census Bureau, "The 2007 statistical abstract," http://www.census.gov/compendia/statab/

47. National Chicken Council press release, 19 May 2004; http://www.news-medical.net/?id=1662

48. National Chicken Council, 2007; http://www.nationalchickencouncil.com/statistics/stat_detail.cfm?id=8

49. C. Li, E.S. Ford, L.C. McGuire, and A.H. Mokdad, "Increasing trends in waist circumference and abdominal obesity among U.S. adults," *Obesity* 15: 216–24 (2007).

50. E.W. Gregg, Y.J. Cheng, B.L. Cadwell, G. Imperatore, D.E. Williams, K.M. Flegal, K.M.V. Narayan, and D.F. Williamson, "Secular trends in cardiovascular disease risk factors according to body mass index in US adults," *JAMA* 293: 1868–74 (2005).

51. Nicholas Stein, "Son of a Chicken Man," *Fortune*, 13 May 2002.

52. Human Rights Watch, *Blood, Sweat, and Fear: Workers' Rights in U.S. Meat and Poultry Plants*, New York: Human Rights Watch, 2005; http://www.hrw.org/reports/2005/usa0105/

53. Eric Schlosser, *Fast Food Nation*, Boston: Houghton Mifflin, 2001.

54. Interview with Dr. Nabil Muhanna, December 2005.

55. H.J. Lipscomb, C.A. Epling, L.A. Pompeii, and J.M. Dement, "Musculoskeletal symptoms among poultry processing workers and a community comparison group: Black women in low-wage jobs in the rural South," *American Journal of Industrial Medicine* 50: 327–38 (2007).
56. Steve Striffler, *Chicken: The Dangerous Transformation of America's Favorite Food*, New Haven, Connecticut: Yale University Press, 2005.
57. Email interview with Steve Striffler, December 2005.
58. S. Striffler, "Inside a poultry processing plant: an ethnographic portrait," *Labor History* 43: 305–13 (2002).
59. Human Rights Watch, *Blood, Sweat, and Fear*.
60. Telephone interview (via translator) with Maria Chavez and Manuel Chavez, December 2005.
61. Interview with Miranda Cady Hallett, June 2007.
62. J.J. Hostynek, E. Patrick, B. Younger, and H.I. Maibach, "Hypochlorite sensitivity in man," *Contact Dermatitis* 20: 32–7 (1989) and M. Medina-Ramón, J.P. Zock, M. Kogevinas, J. Sunyer, Y. Torralba, A. Borrell, F. Burgos, and J.M. Antó, "Asthma, chronic bronchitis, and exposure to irritant agents in occupational domestic cleaning: a nested case-control study," *Occupational and Environmental Medicine* 62: 598–606 (2005).

CHAPTER 6

1. Vaclav Smil, *Enriching the Earth: Fritz Haber, Carl Bosch, and the Transformation of World Food Production*, Cambridge, Mass.: The MIT Press, 2004.
2. T.E. Crews and M.B. Peoples, "Legume versus fertilizer sources of nitrogen: ecological tradeoffs and human needs," *Agriculture, Ecosystems & Environment* 102: 279–97 (2004).
3. Ibid.
4. Julian Darley, *High Noon for Natural Gas: The New Energy Crisis*, White River Junction, Vt.: Chelsea Green Press, 2004.
5. US Energy Information Administration, "U.S. natural gas markets and prospects for the future" (2001); http://www.eia.doe.gov/oiaf/servicerpt/naturalgas/index.html
6. Ma Rong, "The development of fertilizer industry in China," presented to the International Workshop on Economic Policy Reforms and Agricultural Input Markets: Experiences, Lessons, and Challenges, Cape Town, South Africa, 16–20 October 2000.

7. US Dept. of Energy, "How coal gasification power plants work," http://www.fe.doe.gov/programs/powersystems/gasification/howgasificationworks.html

8. Most of the arguments made in this list are following David Roberts' comments in Grist Magazine, "Coal gasification: 'clean coal' or subsidy-hungry boondoggle?," 13 April 2006; http://gristmill.grist.org/story/2006/4/12/173831/909

9. *Christian Science Monitor*, 26 January 2006.

10. Wang Wenshan, "Overview of the development of China's nitrogenous fertilizer industry: some basic experiences," presented to the International Fertilizer Association Technical Conference, Beijing, China, April 2004, http://www.fertilizer.org/ifa/publicat/biblio/biblio04/summary.asp

11. Katja Schumacher and Jayant Sathaye, "India's fertilizer industry: productivity and energy efficiency," Ernest Orlando Lawrence Berkeley National Laboratory publication number LBNL-41846, 1999, http://ies.lbl.gov/iespubs/41846.pdf

12. Ibid.

13. US Dept. of Energy, "Electricity from the wind: wind energy and the natural gas crisis" (2007); http://www.eere.energy.gov/windandhydro/windpoweringamerica/publications.asp

14. Letter from US Senator Tom Harkin to the Senate Committee on Energy and Natural Resources, 5 April 2005; http://harkin.senate.gov/news.cfm?id=236144

15. Wen Huang, "U.S. increasingly imports nitrogen and potash fertilizer," *Amber Waves*, February 2004.

16. One interesting case is Bangladesh, a country that is unlucky in so many ways but does have good natural gas resources. The country uses every cubic foot of gas it can extract, 70 percent of it going to produce either electricity or nitrogen fertilizer. Still, Bangladesh's natural gas consumption per person is only about 4 percent that of the United States. See Ian Blakeley, "Bangladesh—a natural gas perspective," Petroleum Exploration Society of Great Britain (2006); http://energy.ihs.com/News/published-articles/articles/

17. K.F. Isherwood, "The state of the fertilizer industry past, present and future," Sixty-eighth IFA annual conference, Oslo, Norway, 22–25 May 2000; http://www.fertilizer.org/ifa/publicat/PDF/2000_biblio_30.pdf; and Government of India, "Report of the Group on India: Hydrocarbons Vision 2025" (2000); see news report at http://www.thehindubusinessline.com/businessline/2000/03/24/stories/14245004.htm

18. Data obtained from Energy Information Administration, US Department of Energy, http://www.eia.doe.gov/emeu/recs/byfuels/2001/byfuel_ng.pdf

19. Interview with S.P. Wani, February 2007.

20. US Centers for Disease Control and Prevention, "A survey of the quality of water drawn from domestic wells in nine Midwest states," publication NCEH 97-0265 (1998).

21. Peter Weyer, "Nitrate in drinking water and human health," Agriculture Safety and Health Conference, University of Illinois, Urbana-Champaign, Ill., March 2001. http://www.cheec.uiowa.edu/nitrate/health.html

22. F.R. Greer and M. Shannon, "Infant methemoglobinemia: the role of dietary nitrate in food and water," *Pediatrics* 116: 784–6 (2005).

23. P.M. Vitousek, J. Aber, R.W. Howarth, G.E. Likens, P.A. Matson, D.W. Schindler, W.H. Schlesinger, and G.D. Tilman, "Human alteration of the global nitrogen cycle: causes and consequences," *Issues in Ecology* 1: 1–15 (1997).

24. J.N. Galloway and E.B. Cowling, "Reactive nitrogen and the world: 200 years of change," *Ambio* 31: 64–71 (2002).

25. See T.W. Patzek, "Thermodynamics of the corn-ethanol biofuel cycle," *Critical Reviews in Plant Sciences* 23: 519–67 (2004) and T.W. Patzek and D. Pimentel, "Thermodynamics of energy production from biomass," *Critical Reviews in Plant Sciences* 24: 327–64 (2005).

26. James Finch, "Natural gas investors to benefit from global ethanol boom," *Seeking Alpha*, 30 April 2007; http://energy.seekingalpha.com/article/33925

27. Natural Gas Supply Association; http://www.naturalgas.org/overview/background.asp

28. Food and Agriculture Organization, "Voluntary guidelines to support the progressive realization of the right to adequate food in the context of national food security," http://www.fao.org/docrep/meeting/007/j0492e.htm

29. Herman Daly, *Steady State Economics*, Second Edition, Washington, DC: Island Press, 1991, pp. 39–44.

CHAPTER 7

1. William Stanley Jevons, *The Coal Question: An Inquiry Concerning the Progress of the Nation, and the Probable Exhaustion of Our Coal Mines*. London: Macmillan and Co. (Second Edition revised), 1866. First published 1865.

2. I.M. Held, T.L. Delworth, J. Lu, K.L. Findell, and T.R. Knutson, "Simulation of Sahel drought in the 20th and 21st centuries," *Proceedings of the National Academy of Sciences* 102: 17891–96 (2005) and L.D. Rotstayn and U. Lohmann, "Tropical rainfall trends and the indirect aerosol effect," *Journal of Climate* 15: 2103–16 (2002).

3. G. Stanhill and S. Cohen, "Global dimming: a review of the evidence for a widespread and significant reduction in global radiation with discussion of its probable causes and possible agricultural consequences," *Agricultural and Forest Meteorology* 107: 255–78 (2001).

4. P. Alpert, P. Kishcha, Y.J. Kaufman, and R. Schwartzbard, "Global dimming or local dimming?: Effect of urbanization on sunlight availability," *Geophysical Research Letters* 32: L17802, DOI:10.1029/2005GL023320 (2005).

5. V. Ramanathan, C. Chung, D. Kim, T. Bettge, L. Buja, J.T. Kiehl, W.M. Washington, Q. Fu, D.R. Sikka, and M. Wild, "Atmospheric brown clouds: impacts on South Asian climate and hydrological cycle," *Proceedings of the National Academy of Sciences* 102: 5326–33 (2005) and United Nations Environment Program, "The atmospheric brown cloud: climate and other environmental impacts," http://www.rrcap.unep.org/issues/air/impactstudy/index.cfm

6. R. Ramachandran, "Monitoring monsoon," *Frontline* (India), 21 October–3 November 2006.

7. Information in this and the following paragraph is from interviews with Y.V. Malla Reddy, Acción Fraterna, Anantapur, December 2006.

8. Ramanathan et al., "Atmospheric brown clouds."

9. Data from the indispensable http://globalis.gvu.unu.edu

10. Leonard Weiss, "Power points," *Bulletin of the Atomic Scientists*, May/June 2006.

11. Ministry of Commerce and Industry, Government of India, "Fact sheet on foreign direct investment (FDI) from August 1991 to October 2006," http://www.dipp.nic.in/fdi_statistics/india_fdi_nov_2006.pdf

12. Ibid.

13. *Financial Express* (India), 9 January 2007.

14. Embassy of India, Washington, DC; http://www.indianembassy.org/newsite/indoustrade.asp

15. P. Grimes and J. Kentor, "Exporting the greenhouse: foreign capital penetration and CO_2 emissions 1980–1996," *Journal of World-Systems Research* 9: 261–75 (2003).

16. K. Gajwani, R. Kanbur, and X. Zhang, "Patterns of spatial convergence and divergence in India and China," Paper prepared for the Annual Bank Conference on Development Economics, St. Petersburg, Russia, 18–19 January 2006; http://siteresources. worldbank.org/INTDECABC2006/Resources/Xiabo.pdf

17. United Nations Development Programme, "Human development report, 2006," http://hdr.undp.org/hdr2006/report.cfm

18. C. Potera, "Asia's two-stroke engine dilemma," *Environmental Health Perspectives* 112: A613 (2004).

19. I worked in the late 1990s for the non-profit Institute for Rural Health Studies (IRHS) in Hyderabad, India, which ran "barefoot clinics" deep in rural areas. IRHS was virtually the only health-care provider for impoverished people from about 120 villages in the area. Government clinics usually went unstaffed because doctors were off attending to their more lucrative private practices, and corporate medicine was concentrated in the cities. I spoke with IRHS's director Dr. Patricia Bidinger in 2007. She said the situation had changed very little in her region since I had worked there and remains typical of much of India.

20. Ramanathan et al., "Atmospheric brown clouds."

21. National Aeronautics and Space Administration press release; http:// earthobservatory.nasa.gov/Newsroom/MediaAlerts/2006/20060117 21484.html

22. Herman Daly, "Economics in a full world," *Scientific American*, September 2005.

23. Rajat Acharyya, "Trade liberalization, poverty, and income inequality in India," India Resident Mission Policy Brief No. 10, Asian Development Bank; http://www.adb.org/Documents/Papers/ INRM-PolicyBriefs/inrm10.pdf

24. B. Willers, "Sustainable development: a New World deception," *Conservation Biology* 8: 1146–8 (1994).

25. http://globalis.gvu.unu.edu

26. M.E. Moses and J.H. Brown, "Allometry of human fertility and energy use," *Ecology Letters* 6: 295–300 (2003).

27. Herman Daly, "Sustainable development: Definitions, principles, policies." Address to the World Bank, 30 April 2002; http://www. publicpolicy.umd.edu/faculty/daly/World%20Bank%20speech%2 0com%202.pdf

28. Take the difference between US and Indian per capita ecological footprint and multiply by a population of 1.4 billion, which India will soon have or may already have; that total additional footprint is similar to that of humanity as a whole. Figures are from Redefining Progress, "2005 Ecological Footprint of Nations," http://www.

rprogress.org/publications/2006/Footprint%20of%20Nations%2
02005.pdf

29. P. Crutzen, "Albedo enhancement by stratospheric sulfur injections: a contribution to resolve a policy dilemma?," *Climatic Change* 77: 211–20 (2006).

30. B. Govindasamy and K. Caldeira, "Geoengineering Earth's radiation balance to mitigate CO_2-induced climate change," *Geophysical Research Letters* 27: 2141–4 (2000).

CHAPTER 8

1. Karl Marx, *Capital: an Abridged Edition*, David McClellan (ed.), Oxford, UK: Oxford University Press, 1995, p. 321.

2. Field Maloney, "Is Whole Foods wholesome? The dark secrets of the organic-food movement," *Slate*, 17 March 2006; http://www.slate.com/id/2138176

3. The *Independent*, 12 January 2006.

4. The *Times* of London, 2 August 2007.

5. "The 100 best companies to work for, 2007," *Fortune* (undated); http://money.cnn.com/magazines/fortune/bestcompanies/2007/snapshots/5.html

6. Pamela Danziger, *Why People Buy Things They Don't Need: Understanding and Predicting Consumer Behavior*, Chicago: Dearborn Trade Publishing, 2004, p. 23.

7. Seth Lubove, "Food Porn," *Forbes*, 14 February 2005.

8. *Wall Street Journal*, 4 December 2006.

9. http://money.cnn.com/2005/10/25/news/fortune500/walmart_wage

10. *Fayetteville* [Arkansas] *Morning News*, 2 June 2007.

11. Stan Cox, "Wal-Mart wages don't support Wal-Mart workers," *AlterNet*, 10 June 2003; http://www.alternet.org/story/16111

12. Stan Cox, "Natural food, unnatural prices," *AlterNet*, 25 January 2006; http://www.alternet.org/story/31260

13. Telephone interview with Ashley Hawkins, January 2006.

14. *Wall Street Journal*, 4 December 2006.

15. Hannah Clark, "Whole Foods: spinning CEO pay," *Forbes*, 20 April 2006.

16. John Brewer, "Half truths at Whole Foods," *San Antonio Current*, 25 July 2002.

17. Email interview with Jeremy Plague, January 2006.

18. *Wall Street Journal*, 4 December 2006.

19. Mary Holz-Clause and Malinda Geisler, "Grocery retailing profile," Agricultural Marketing Resource Center, September 2006; http://www.agmrc.org/agmrc/markets/Food/groceryindustry.htm
20. *Fortune*, "The 100 best companies to work for, 2007."
21. "Austinist interviews Michael Pollan, author of The Omnivore's Dilemma," *Austinist*, 25 May 2006; http://www.austinist.com/archives/2006/05/25/austinist_interviews_michael_pollan_author_of_the_omnivores_dilemma.php
22. Maloney, "Is Whole Foods wholesome?"
23. Michael Pollan, *The Omnivore's Dilemma*, New York: Penguin, 2006.
24. http://www.michaelpollan.com/article.php?id=80 and http://www.wholefoods.com/blogs/jm/archives/2006/06/detailed_reply.html
25. Ibid.
26. Pallavi Gogoi, "Are Wal-Mart's 'organics' organic?," *Business Week*, 18 January 2007.
27. Pallavi Gogoi, "Wal-Mart's organic offensive," *Business Week*, 29 March 2007.
28. *Wall Street Journal*, 4 December 2006.
29. Email interview with Rhonda Janke, January 2006.
30. Telephone interview with Brahm Ahmadi, January 2006.
31. Land area calculations were based on figures from the UN Food and Agriculture Organization, http://faostat.fao.org
32. http://www.michaelpollan.com/article.php?id=80
33. Jerry D. Glover, Cindy M. Cox, and John P. Reganold, "Future farming: a return to roots?," *Scientific American*, August 2007.
34. T.S. Cox, J.D. Glover, D.L. Van Tassel, C.M. Cox, and L.R. DeHaan, "Prospects for developing perennial grain crops," *Bioscience* 56: 649–59 (2006).
35. Lubove, "Food Porn."
36. T.S. Cox, M. Bender, C. Picone, D.L. Van Tassel, J.B. Holland, E.C. Brummer, B.E. Zoeller, A.H. Paterson, and W. Jackson, "Breeding perennial grain crops," *Critical Reviews in Plant Sciences* 21: 59–91 (2002).
37. http://www.amazinggrains.com
38. Lubove, "Food Porn."
39. *Appliance* magazine, March 2007.
40. Associated Press, 31 March 2007.

CHAPTER 9

1. DuPont's "Heritage" website; http://heritage.dupont.com/touchpoints/tp_1939/depth.shtml

2. T.H. Begley, K. White, P. Honigfort, M. Twaroski, R. Neches, and R.A. Walker, "Perfluorochemicals: potential sources of and migration from food packaging," *Food Additives and Contaminants* 22: 1023–103 (2005); EPA, "Revised draft hazard assessment of perfluorooctanoic acid and its salts," 4 November 2002. Copy may be viewed at http://www.ewg.org/issues_content/pfcs/20021113/pdf/EPA_PFOA_110402.pdf (last accessed 10 June 2007); 3M Company, "Biodegradation study report, revision 1: Biodegradation screen study for telomer-type alcohols," 6 November 2002. Copy may be viewed at http://www.ewg.org/reports/pfcworld/pdf/sludge_full.pdf (last accessed 10 June 2007).

3. N. Yamashita, K. Kannan, S. Taniyasu, Y. Horii, G. Petrick, and T. Gamo, "A global survey of perfluorinated acids in oceans," *Marine Pollution Bulletin* 51: 658–68 (2005); M. Houde, T.A. Bujas, J. Small, R.S. Wells, P.A. Fair, G.D. Bossart, K.R. Solomon, and D.C. Muir, "Biomagnification of perfluoroalkyl compounds in the bottlenose dolphin (*Tursiops truncatus*) food web," *Environmental Science and Technology* 40(13): 4138–44 (2006); K. Kannan, S. Corsolini, J. Falandysz, G. Oehme, S. Focardi, and J.P. Giesy, "Perfluoroctanesulfonate and related fluorinated hydrocarbons in marine mammals, fishes and birds from coasts of the Baltic and Mediterranean Seas," *Environmental Science and Technology* 36: 3210–16 (2002); K.I. Van de Vijver, P. Hoff, K. Das, S. Brasseur, W. Van Dongen, E. Esmans, P. Reijnders, R. Blust, and W. De Coen, "Tissue distribution of perfluorinated chemicals in harbor seals (*Phoca vitulina*) from the Dutch Wadden Sea," *Environmental Science and Technology* 39: 6978–84 (2005); J.M. Keller, K. Kannan, S. Taniyasu, N. Yamashita, R.D. Day, M.D. Arendt, A.L. Segars, and J.R. Kucklick, "Perfluorinated compounds in the plasma of loggerhead and Kemp's Ridley sea turtles from the southeastern coast of the United States," *Environmental Science and Technology* 39: 9101–8 (2005); R. Bossia, F.F. Rigeta, R. Dietza, C. Sonnea, P. Fausera, M. Damb, and K. Vorkampa, "Preliminary screening of perfluorooctane sulfonate (PFOS) and other fluorochemicals in fish, birds and marine mammals from Greenland and the Faroe Islands," *Environmental Pollution* 136: 323–9 (2005); L. Tao, K. Kannan, N. Kajiwara, M.M. Costa, G. Fillman, S. Takahashi, and S. Tanabe, "Perfluorooctanesulfonate and related fluorochemicals in albatrosses, elephant seals, penguins, and polar skuas from the Southern Ocean," *Environmental Science and Technology* 40: 7642–8 (2006); K. Kannan, J. Newsted, R.S. Halbrook, and J.P. Giesy, "Perfluorooctanesulfonate and related fluorinated hydrocarbons in mink and river otters from the United States,"

Environmental Science and Technology 36: 2566–71 (2002); K. Kannan, J.C. Franson, W.W. Bowerman, K.J. Hansen, P.D. Jones, and J.P. Giesy, "Perfluorooctane sulfonate in fish-eating water birds including bald eagles and albatrosses," *Environmental Science and Technology* 35: 3065–70 (2001); M. Smithwick, R.J. Norstrom, S.A. Mabury, K. Solomon, T.J. Evans, I. Stirling, M.K. Taylor, and D.C. Muir, "Temporal trends of perfluoroalkyl contaminants in polar bears (*Ursus maritimus*) from two locations in the North American Arctic, 1972–2002," *Environmental Science and Technology* 40: 1139–43 (2006); K.E. Holmstrom, U. Jarnberg, and A. Bignert, "Temporal trends of PFOS and PFOA in guillemot eggs from the Baltic Sea, 1968–2003," *Environmental Science and Technology* 39: 80–4 (2005); and K. Kannan, L. Tao, E. Sinclair, S.D. Pastva, D.J. Jude, and J.P. Giesy, "Perfluorinated compounds in aquatic organisms at various trophic levels in a Great Lakes food chain," *Archives of Environmental Contamination and Toxicology* 48: 559–66 (2005).

4. J. Falandysz, S. Taniyasu, A. Gulkowska, N. Yamashita, and U. Schulte-Oehlmann, "Is fish a major source of fluorinated surfactants and repellents in humans living on the Baltic coast?," *Environmental Science and Technology* 40: 748–51 (2006).

5. K. Kannan, S. Corsolini, J. Falandysz, G. Fillmann, K.S. Kumar, B.G. Loganathan, M. Ali Mohammed, J. Olivero, N. Van Wouwe, Jae Ho Yang, and K.M. Aldous, "Perfluorooctanesulfonate and related fluorochemicals in human blood from several countries," *Environmental Science and Technology* 38: 4489–95 (2004); A.M. Calafat, Z. Kuklenyik, S.P. Caudill, J.A. Reidy, and L.L. Needham, "Perfluorochemicals in Pooled Serum Samples from United States Residents in 2001 and 2002," *Environmental Science and Technology* 40: 2128–34 (2006); G.W. Olsen, T.R. Church, J.P. Miller, J.M. Burris, K.J. Hansen, J.K. Lundberg, J.B. Armitage, R.M. Herron, Z. Medhdizadehkashi, J.B. Nobiletti, E.M. O'Neill, J.H. Mandel, and L.R. Zobel, "Perfluorooctanesulfonate and other fluorochemicals in the serum of American Red Cross adult blood donors," *Environmental Health Perspectives* 111: 1892–901 (2003); G.W. Olsen, H.Y. Huang, K.J. Helzlsouer, K.J Hansen, J.L. Butenhoff, and J.H. Mandel, "Historical comparison of perfluorooctanesulfonate, perfluorooctanoate, and other fluorochemicals in human blood," *Environmental Health Perspectives* 113: 539–45 (2005); and K. Harada, N. Saito, K. Sasaki, K. Inoue, and A. Koizumi, "Perfluorooctane sulfonate contamination of drinking water in the Tama River, Japan: estimated effects in resident serum levels," *Bulletin of Environmental Contamination Toxicology* 71: 31–6 (2003).

6. 3M Company study, 2002. Copy may be viewed at http://www.ewg. org/reports/pfcworld/pdf/kid_blood_full.pdf (last accessed 10 June 2007).

7. K.S. Guruge., S. Taniyasub, N. Yamashitab, S. Wijeratnac, K.M. Mohottid, H.R. Seneviratnec, K. Kannan, N. Yamanakaa and S. Miyazaki, "Perfluorinated organic compounds in human blood serum and seminal plasma: a study of urban and rural tea worker populations in Sri Lanka," *Journal of Environmental Monitoring* 7: 371–7 (2005). The web version at http://www.rsc.org/ej/EM/2005/ b412532k/ includes references to several other studies from the United States, India, Italy, and Japan.

8. Centers for Disease Control and Prevention, "Third national report on human exposure to environmental chemicals, 2005," http://www. cdc.gov/exposurereport/

9. D. Kriebel, J. Tickner, P. Epstein, J. Lemons, R. Levins, E.L. Loechler, M. Quinn, R. Rudel, T. Schettler, and M. Stoto, "The precautionary principle in environmental science," *Environmental Health Perspectives* 109: 871–6 (2001).

10. Meisan Lim, "Teflon's sticky question: Is it bad for you?," Columbia News Service, 31 October 2006; http://jscms.jrn.columbia.edu/ cns/2006-10-31/lim-howbadisteflon

11. D.E. Bowers, "Cooking trends echo changing roles of women," *Food Review* 23(1): 23–9 (2000).

12. H. Hartwell, "Catering for health: a review," *Journal of the Royal Society for the Promotion of Health* 125: 113–16 (2005).

13. W.A. Temple, I.R. Edwards, and S.J. Bell, "'Poly' fume fever—two fatal cases," *New Zealand Veterinary Journal* 33: 30 (1985). Many more incidents cited at http://www.ewg.org/reports/toxicteflon/ toxictemps.php (last accessed 10 June 2007).

14. C.H. Lee, Y.L. Guo, P.J. Tsai, H.Y. Chang, C.R. Chen, C.W. Chen, and T.R. Hsiue, "Fatal acute pulmonary oedema after inhalation of fumes from polytetrafluoroethylene (PTFE)," *European Respiratory Journal* 10: 1408–11 (1997).

15. M. Boucher, T.J. Ehmler, and A.J. Bermudez, "Polytetrafluoroethyl-ene gas intoxication in broiler chickens," *Avian Diseases* 44: 449–53 (2000).

16. D. Bracco and J.B. Favre, "Pulmonary injury after ski wax inhalation exposure," *Annals of Emergency Medicine* 32: 616–19 (1998).

17. "Nonstick pans are OK in new tests," *Consumer Reports*, June 2007.

18. B. Jugg, J. Jenner, and P. Rice, "The effect of perfluoroisobutene and phosgene on rat lavage fluid surfactant phospholipids," *Human and Experimental Toxicology* 18: 659–68 (1999) and H. Arito and R. Soda, "Pyrolysis products of polytetrafluoroethylene and

polyfluoroethylenepropylene with reference to inhalation toxicity," *Annals of Occupational Hygiene* 20: 247–53 (1977).

19. D.A. Ellis, S.A. Mabury, J.W. Martin, and D.C.G. Muir, "Thermolysis of fluoropolymers as a potential source of halogenated organic acids in the environment," *Nature* 412: 321–4 (2001).

20. K. Johns and G. Stead, "Fluoroproducts—the extremophiles," *Journal of Fluorine Chemistry* 104: 5–18 (2000).

21. http://www.teflon.com/Teflon/teflonissafe/keyquestions.html

22. Lim, "Teflon's sticky question."

23. C. Lau, J.L. Butenhoff, and J.M. Rogers, "The developmental toxicity of perfluoroalkyl acids and their derivatives," *Toxicology and Applied Pharmacology* 198: 231–41 (2004); C. Lau, J.R. Thibodeaux, R.G. Hanson, M.G. Narotsky, J.M. Rogers, A.B. Lindstrom, and M.J. Strynar, "Effects of perfluorooctanoic acid exposure during pregnancy in the mouse," *Toxicological Science* 90: 510–18 (2006); S.S. White, A.M. Calafat, Z. Kuklenyik, L. Villanueva, R.D. Zehr, L. Helfant, M.J. Strynar, A.B. Lindstrom, J.R. Thibodeaux, C. Wood, and S.E. Fenton, "Gestational PFOA exposure of mice is associated with altered mammary gland development in dams and female offspring," *Toxicological Sciences* 96: 133–44 (2007); C.J. Wolf, S.E. Fenton, J.E. Schmid, A.M. Calafat, Z. Kuklenyik, X.A. Bryant, J. Thibodeaux, K.P. Das, S.S. White, C.S. Lau, and B.D. Abbott, "Developmental toxicity of perfluorooctanoic acid in the cd-1 mouse after cross-foster and restricted gestational exposures," *Toxicological Science* 95: 462–73 (2007); Raymond G. York, "Oral (Gavage) two generation (one litter per generation) reproduction study of ammonium perfluorooctonate (AFPO) in rats," Sponsor's Study Number T-6889.6, submitted to EPA, 26 March 2002.

24. White et al., "Gestational PFOA exposure."

25. Q. Yang, Y. Xie, A.M. Eriksson, B.D. Nelson, and J.W. DePierre, "Further evidence for the involvement of inhibition of cell proliferation and development in thymic and splenic atrophy induced by the peroxisome proliferator perfluoroctanoic acid in mice," *Biochemical Pharmacology* 62: 1133–40 (2001).

26. R. Nilsson, B. Beije, V. Préat, K. Erixon, and C. Ramel, "On the mechanism of the hepatocarcinogenicity of peroxisome proliferators," *Chemico-Biological Interactions* 78: 235–50 (1991).

27. G.L. Kennedy, Jr., J.L. Butenhoff, G.W. Olsen, J.C. O'Connor, A.M. Seacat, R.G. Perkins, L.B. Biegel, S.R. Murphy, and D.G. Farrar, "The toxicology of perfluorooctanoate," *Critical Reviews in Toxicology* 34: 351–84 (2004) and EPA, "Revised draft hazard assessment."

28. J.L. Butenhoff, G.L. Kennedy, Jr., S.R. Frame, S.R., J.C. O'Connor, and R.G. York, "The reproductive toxicology of ammonium per-

fluorooctanoate (APFO) in the rat," *Toxicology* 196: 95–116 (2004).

29. Kennedy et al., "The toxicology of perfluorooctanoate."
30. http://www2.dupont.com/PFOA/en_US/about_pfoa/pfoa_facts. html
31. D Brooke, A. Footitt, and T.A. Nwaogu, "Environmental risk evaluation report: perfluorooctanesulphonate (PFOS)," UK Environment Agency (2004); http://www.fluorideaction.org/ pesticides/pfos.uk.report.2004.pdf and Lau et al., "The developmental toxicity of perfluoroalkyl acids."
32. DuPont news release, 30 January 2006, http://www2.dupont.com/ Media_Center/en_US/news_releases/2006/article20060130b.html
33. EWG, "Overview of worker studies," http://www.ewg.org/reports/ pfcworld/pdf/workertable.pdf (last accessed 10 June 2007).
34. Kennedy et al., "The toxicology of perfluorooctanoate."
35. DuPont company memo, 21 May 1984, which can be viewed at http://www.ewg.org/issues_content/PFCs/20030606/pdf/dupont_ elim_PFOA_1984.pdf (last accessed 10 June 2007).
36. http://www.ewg.org/issues/PFCs/20030606/index.php (last accessed 10 June 2007).
37. Amy Cortese, "DuPont's Teflon dilemma: how Chad Holliday, the champion of sustainability, is managing an environmental challenge," *Chief Executive*, November 2003.
38. DuPont press release, 17 January 2005; http://www2.dupont.com/ PFOA/en_US/pdf/FayettevilleStatement_1_17_06.pdf
39. http://pubs.acs.org/subscribe/journals/esthag-w/2006/feb/business/ figures/DuPont_shareholders_report.pdf?sessid=6006l3
40. C-8 Health Project, http://www.c8healthproject.org/health.htm
41. *Charleston Gazette*, 1 May 2007.
42. *Charleston Gazette*, 19 October 2006 and United Steelworkers press release, http://www.dupontsafetyrevealed.org/Bogus_Study.htm
43. Associated Press, 14 December 2005.
44. *Washington Post*, 29 June 2005 and EPA Science Advisory Board, 30 May 2006: http://www.epa.gov/sab/pdf/sab_06_006.pdf
45. Associated Press, 19 April 2006.
46. KCCI-TV, Des Moines, 19 April 2006; http://www.kcci.com/ news/8834016/detail.html
47. Colin Rigley, "Slick compound targeted as possible carcinogen," *Capitol Weekly* (Sacramento, Calif.), 23 November 2006.
48. Glenn Hess, "PFOA drinking water standard lowered," *Chemical & Engineering News*, 22 November 2006.
49. EPA, "Fact Sheet: EPA, DuPont agree on measures to protect drinking water near the DuPont Washington Works," http://www. epa.gov/region03/enforcement/dupont_factsheet.html

50. EPA press release, 25 January 2006; http://yosemite.epa.gov/opa/admpress.nsf/177f410e8a398c0f85257021005643a7/fd1cb3a075 697aa485257101006afbb9!OpenDocument
51. Associated Press, 5 February 2007.
52. Telephone interview with Kristan Markey, February 2007.
53. Lisa Makson, "Rachel Carson's ecological genocide," *FrontPage* magazine, 31 July 2003, http://www.frontpagemag.com/Articles/ReadArticle.asp?ID=9169
54. F. Powledge, "Millennium Ecosystem Assessment," *BioScience* 56: 880–6 (2006).
55. H. Sanderson, T.M. Boudreau, S.A. Mabury, and K.R. Solomon, "Impact of perfluorooctanoic acid on the structure of the zooplankton community in indoor microcosms," *Aquatic Toxicology* 62: 227–34 (2003).
56. South Mississippi *Sun Herald*, 20 September 2006.
57. *Sun Herald*, 24 September 2006.
58. Telephone interview with Brenda Songy, December 2006.
59. *Sun Herald*, 24 September 2006.
60. US Census Bureau, http://www.census.gov/hhes/www/saipe
61. Environmental Justice and Health Union, "Environmental exposure and racial disparities," August 2003, http://www.ejhu.org/disparities.html
62. Ike Brannon, "What is a life worth?," *Regulation*, 1 December 2004.
63. J. Kaiser, "How much are human lives and health worth?," *Science* 299: 1836–7 (2003).
64. http://www.defendingscience.org/case_studies/perfluorooctanoic-acid.cfm
65. Ft. Worth *Star-Telegram*, 3 December 2006.
66. Ibid.
67. Government Accountability Office, "Chemical regulation: options exist to improve EPA's ability to assess health risks and manage its chemical review program," publication GAO-05-458, June 2005; http://www.gao.gov/new.items/d05458.pdf
68. K.C. Lowe, "Blood substitutes: from chemistry to clinic," *Journal of Materials Chemistry* 16: 4189–96 (2006).

CHAPTER 10

1. Daniel Esty and Andrew Winston, *Green to Gold: How Smart Companies Use Environmental Strategy to Innovate, Create Value, and Build Competitive Advantage*, New Haven, Connecticut: Yale University Press, 2006.

2. The *Independent*, 12 June 2005.
3. *Chicago Tribune*, 30 July 2007. The following month, responding to widespread outrage, BP officials pledged not to take advantage of the new, higher pollution limits but warned that the company might have to scuttle plans for expansion of its Indiana facilities. Northwest Indiana's *Post Tribune* (2 September 2007), having discovered that BP had been caught exceeding the old pollution limits ten times in the previous nine years, noted that the new pledge "doesn't mean the company won't exceed those limits" in the future.
4. Jon Entine, "The stranger-than-truth story of the Body Shop," in *Killed: Great Journalism too Hot to Print*, David Wallis (ed.), New York: Nation Books, 2004.
5. Terence Turner, "Neoliberal ecopolitics and indigenous peoples: the Kayapo, the 'Rainforest Harvest,' and The Body Shop," *Yale F & ES Bulletin* 98: 113–27.
6. *Ethical Consumer* magazine (Manchester, UK) press release, 17 March 2006; http://www.ethicalconsumer.org/bodyshop_loreal. htm
7. Karl Marx, *Capital: an Abridged Edition*, David McClellan (ed.), Oxford, UK: Oxford University Press, 1995, p. 298.
8. Nicholas Georgescu-Roegen, *The Entropy Law and the Economic Process*, New York: Harvard University Press, 1971, pp. 250–1.
9. William Stanley Jevons, *The Coal Question: An Inquiry Concerning the Progress of the Nation, and the Probable Exhaustion of Our Coal Mines*, London: Macmillan and Co. (Second Edition revised), 1866. First published 1865; e-book retrieved from http://www. eoearth.org/article/The_Coal_Question_
10. Marx, *Capital*, pp. 64–100.
11. Ibid., p. 98.
12. This view has been elaborated most thoroughly by David Korten in books such as *When Corporations Rule the World* (Second Edition, San Francisco: Berrett-Koehler Publishers, 2001) and in the magazine *Yes!*, of which he was a founder. But Korten doesn't explain how smaller-scale, more humane capitalism (which, he maintains, isn't even capitalism), can be achieved without addressing class exploitation and the M–C–M' growth cycle. He has even used the Visa International credit-card and Ace Hardware, Inc. as models for transcending capitalism (Korten, *The Post-Corporate World: Life After Capitalism*, San Francisco: Berrett-Koehler Publishers, 2000, pp. 176–8.) Ecosocialist Joel Kovel has written that "Korten has no difficulty in seeing [the capitalist state] checked by 'global civilizing society,' which will restrain and effectively domesticate the animal, leading to the Neo-Smithian Promised Land. This is essentially an

upbeat fairy tale standing in for history, and if it were true, the world would be a much easier place to change." (*The Enemy of Nature: The End of Capitalism or the End of the World?*, New York: Zed Books, 2002, p. 162.)

13. Marx, *Capital*, p. 379.

14. Ibid., p. 347.

15. Georgescu-Roegen, *The Entropy Law and the Economic Process*, p. 304.

16. See D.M. Levy, "How the dismal science got its name: debating racial quackery," *Journal of the History of Economic Thought* 23: 5–35 (2001).

17. Herman Daly, *Steady State Economics*, Washington, DC: Island Press, 1991, pp. 50–76.

18. Herman Daly and Joshua Farley, *Ecological Economics: Principles and Applications*, Washington, DC: Island Press, 2004.

19. Jevons, *The Coal Question*, VII.25–6.

20. Ibid., VII.10.

21. Ibid., VII.3.

22. Herman Daly, "Sustainable development: Definitions, principles, policies," Address to the World Bank, 30 April 2002; http://www. publicpolicy.umd.edu/faculty/daly/World%20Bank%20speech%2 0com%202.pdf

23. Ibid.

24. Arguments for blunting the effects of growth through improved efficiency and owners' goodwill are exemplified by the widely read book *Natural Capitalism* (Paul Hawken, Amory B. Lovins, and L. Hunter Lovins, *Natural Capitalism*, New York: Little, Brown, 1999). When, in a 2004 email interview, I asked lead author Paul Hawken if he thinks Jevons's efficiency trap still applies in the twenty-first century, he sidestepped the issue by claiming that near-perfect efficiency is achievable (presumably allowing near-infinite growth). "Amory and I fully believe," he wrote, "that a 99 percent reduction in the throughput of energy and resources is possible and will eventually occur." Stan Cox, "From here to economy," *Grist Magazine*, 23 April 2004; http://grist-staging.electricembers.net/ news/maindish/2004/04/23/cox-economy/index.html

25. US Energy Information Administration, November 2006; http:// www.eia.doe.gov/oiaf/1605/ggrpt/carbon.html

26. Basel Action Network and Silicon Valley Toxics Coalition, "Exporting harm: the high-tech trashing of Asia," 25 February 2002; http:// www.svtc.org/cleancc/pubs/technotrash.pdf

27. J. Perraton, "Heavy constraints on a weightless world," *American Journal of Economics and Sociology* 65: 641–91 (2006).

28. E.C. Alfredsson, "'Green' consumption—no solution for climate change," *Energy* 29: 513–24 (2004).

29. *USA Today*, 19 September 2006.

30. Travel Industry Association of America, "2004 Business Travelers' Survey," http://www.tia.org/Pubs/pubs.asp?PublicationID=92

31. National Business Travel Association press release, 30 November 2006; www.nbta.org/About/News/Releases2006/rls112006.htm

32. Carson Wagonlit Travel press release, 23 January 2006.

33. British scientists and government officials cited in the *Independent*, 25 May 2005.

34. *USA Today*, 19 September 2006.

35. *USA Today*, 7 September 2004.

36. Jeremy Rifkin, *The European Dream: How Europe's Vision of the Future is Quietly Eclipsing the American Dream*, New York: Tarcher/Penguin, 2004.

37. Steve McGiffen, "A really bad book by someone who should know better," *Spectrezine*, 6 December 2005; http://www.spectrezine.org/reviews/McGiffen.htm

38. *Fortune*, "Europe's top 50," http://money.cnn.com/magazines/fortune/global500/2006/europe/

39. ETC Group, "Global seed industry concentration—2005," http://www.etcgroup.org/upload/publication/pdf_file/48

40. T. Lang, G. Rayner, and E. Kaelin, "The food industry, diet, physical activity and health: a review of reported commitments and practice of 25 of the world's largest food companies, measured against the goals of the World Health Organisation global strategy on diet, physical activity and health," Centre for Food Policy, City University London (2006); http://www.city.ac.uk/press/The%20Food%20Industry%20Diet%20Physical%20Activity%20and%20Health.pdf

41. Canadian Broadcasting Corporation, 3 February 2003; http://www.cbc.ca/news/features/water/business.html

42. *Sunday Telegraph*, 2 July 2006.

43. Mark Rainer, "Kyoto's Clean Development Mechanism: global warming and its market fix," *World Socialist Website*, 13 January 2007; http://www.wsws.org/articles/2007/jan2007/glob-j13.shtml

44. Stan Cox, "From here to economy."

45. Jonathan Ansfield, "Beijing battles to control its booming coal biz," *Newsweek International*, 15 January 2007.

46. Jared Diamond, *Collapse: How Societies Choose to Fail or Succeed*, New York: Viking, 2005.

47. R. Smith, "The engine of eco *Collapse*," *Capitalism Nature Socialism* 16: 19–36 (2005).

48. Herman Daly, "Economics in a full world," *Scientific American*, September 2005.
49. Joel Kovel, *The Enemy of Nature: The End of Capitalism or the End of the World?*, New York: Zed Books, 2002.
50. John Bellamy Foster, *Ecology Against Capitalism*, New York: Monthly Review Press, 2002.
51. John Bellamy Foster, *The Vulnerable Planet*, New York: Monthly Review Press, 1993.
52. Michael Löwy, "What is ecosocialism?," *Capitalism Nature Socialism* 16(2): 15–24.
53. Lawrence Mishel, Jared Bernstein, and Sylvia Allegretto, *The State of Working America 2006/2007*, Washington, DC: Economic Policy Institute, 2006.
54. Marx, *Capital*, p. 362.

SUGGESTED READING

These books include novels as well as book-length essays and quasi-textbooks, and while some are heavier going than others, they are all accessible to the non-expert. I've listed them in chronological order.

The Jungle by Upton Sinclair (1906). A book still without equal in social fiction, *The Jungle* follows an immigrant family through the horrors of Chicago's turn-of-the-century meatpacking industry. Sinclair's book prompted passage of meat-inspection laws, but contamination remains a problem and the exploitation of meat workers persists today.

The Grapes of Wrath by John Steinbeck (1939). The monumental novel showing how Nature may have provided the drought but political and economic forces created the Dust Bowl and the Great Depression.

Silent Spring by Rachel Carson (Houghton Mifflin, 1962). Written in straightforward and captivating language by a marine biologist, this was the document that marked the beginning of the modern environmental movement. Because of it, pesticides were banned on the basis of broad ecological thinking, not just human disease.

The Entropy Law and the Economic Process by Nicholas Georgescu-Roegen (Harvard University Press, 1971). The book showing that there's no free lunch, that we can only work to moderate our damage to the planet as a whole; we cannot have a neutral or positive impact. Part physics, part mathematics, part philosophy, part biology, and all economics, this is not a book to take to the beach (although I once unwisely did that). Even when the details are difficult, the arguments are strong and convincing. The final two chapters can almost be read alone. They are almost poetic, but with some equations.

The Unsettling of America: Culture and Agriculture by Wendell Berry (Sierra Club Books, 1977). The book that brought back agrarianism, by the renowned novelist, poet, and essayist, vividly illustrated the industrialization of agriculture and destruction of the family farm. Twenty-five years after the book's publication, as it was being hailed as prophetic, Berry noted sadly that he had meant it as a warning, not a prophecy.

New Roots for Agriculture by Wes Jackson (University of Nebraska Press, 1980). Jackson's classic book demonstrated that agriculture as we know it, featuring the plow and annual crops, has been destroying soil not just since the Dust Bowl but for the past 10,000 years, and that only Nature, not the Market, holds the answer.

Steady State Economics, Second Edition by Herman E. Daly (Island Press, 1991). Daly organized and extended the arguments made in *The Entropy Law and the Economic Process* to show how the human economy can and must be curtailed and squeezed inside the ecosphere, and not the other way around.

The Tortilla Curtain by T. Coraghessan Boyle (Viking, 1995). A gripping novel focused on an immigrant couple from Mexico vainly seeking work, food, and shelter in the smug suburbs of southern California.

Divided Planet: The Ecology of Rich and Poor by Tom Athanasiou (Little, Brown, 1996). A stark lesson in how "the environment" can look very different depending on one's economic status—and how that's not a result of random chance.

Marx's Ecology: Materialism and Nature by John Bellamy Foster (Monthly Review Press, 2000). The mainstream environmental movement has never revered Karl Marx; indeed, he's been regarded as an antagonist who championed the human domination of Nature. But Foster argues that Marx keenly analyzed humanity's impact on the natural world, especially when it came to the nineteenth century's number-one ecological concern: the degradation of agricultural soils.

Stolen Harvest: The Hijacking of the Global Food Supply by Vandana Shiva (Zed Books, 2001). Shiva has become the global South's standard-bearer in its battle with seed, biotech, and pharmaceutical firms seeking to get a tight grip on the crop gene pool.

The Constant Gardener by John le Carré (Hodder & Stoughton, 2001). The master of the spy novel turns our attention to pharmaceutical companies' crimes against humanity in Africa.

Nickel and Dimed: On (Not) Getting by in America by Barbara Ehrenreich (Henry Holt & Co., 2001). A highly regarded writer from the left goes to work in a series of low-wage jobs herself and shows us the damage it can do. What Wal-Mart really means by its slogan "Our People Make the Difference."

Everybody Loves a Good Drought: Stories from India's Poorest Districts by P. Sainath (Penguin India, 2002). In unadorned language that's almost impossible to stop reading until the book's over, this veteran newspaper reporter reveals what it's really like to live in the country's impoverished countryside. We see rapacious moneylenders and heartless government officials on one side and endlessly resourceful citizens on the other. Nobel prizewinner Amartya Sen called him "one of the world's great experts on famine and hunger."

Fast Food Nation by Eric Schlosser (Houghton Mifflin, 2002). In this runaway bestseller, the economy treats everyone under the Golden

Arches—the customer, the employee, the meat worker, and the farmer—almost as badly as is the feedlot steer.

The Enemy of Nature: The End of Capitalism or the End of the World? by Joel Kovel (Zed Books, 2002). Kovel makes the fundamental arguments showing how the addiction to growth means that we can have capitalism or a livable planet but not both. His dissection of philosophies and movements that attempt to avoid or resolve that contradiction is especially revealing.

After Capitalism by David Schweikart (Rowman & Littlefield, 2002). An attempt to envision a society based on worker ownership and public investment. Schweikart's explanation of capitalism's fatal flaws is one of the clearest around, for those of us who aren't economists. His outline of a market-socialist economy and his brief explanation of how it could be more ecologically sustainable are unconvincing.

The Perverse Economy: The Impact of Markets on People and the Environment by Michael Perelman (Palgrave Macmillan, 2003). Starting and ending with what he refers to as Adam Smith's "farm worker paradox"—that people who do the hardest work in society, using their skill and knowledge to provide the most important necessities of life, are paid the least—Perelman provides another clear explanation of capitalist economics.

Coronary Artery Disease: Genes, Drugs, and the Agricultural Connection by Ole Faergerman (Elsevier, 2003). A Danish medical expert explains how a disease with multiple, complex causes is a product of society and cannot be "cured" with drugs. The dairy industry plays the role of chief villain.

Critical Condition: How Health Care in America Became Big Business & Bad Medicine by Donald Bartlett and James Steele (Doubleday, 2004). This pair of award-winning newspaper reporters occasionally turns out devastating investigative books as well, and this one on Big Medicine is superb.

Selling Sickness: How The World's Biggest Pharmaceutical Companies Are Turning Us All Into Patients by Ray Moynihan and Alan Cassels (Greystone Books, 2005). Disease mongering explained, in all its shocking forms.

The Long Emergency: Surviving the Converging Catastrophes of the Twenty-First Century by James Howard Kunstler (Grove Press, 2006). Kunstler paints a frightening picture of the future, putting in practical terms Nicholas Georgescu-Roegen's prediction that "the price of technological progress has meant a shift from the more abundant source of low entropy—the solar radiation—to the less abundant one—the earth's mineral resources ... the destiny of the human species is to choose a truly great but brief, not a long and dull, career."

The Omnivore's Dilemma: A Natural History of Four Meals by Michael Pollan (Penguin, 2006). By following those meals, ranging from synthetic/industrial to self-scavenged, back to their sources, Pollan has helped open America's eyes to what's in our stomachs.

Animal, Vegetable, Miracle: A Year of Food Life by Barbara Kingsolver with Steven L. Hopp and Camille Kingsolver (HarperCollins, 2007). In the words of the *New York Times*, Kingsolver "expresses the basic tenets of Slow Food International and sustainable agriculture ... But she succeeds in dramatizing her own family's story so that these ideas come to life, anecdotally and charmingly."

INDEX

Compiled by Kathleen League

213